T0140001

Computer Communications and Networks

The **Computer Communications and Networks** series is a range of textbooks, monographs and handbooks. It sets out to provide students, researchers, and non-specialists alike with a sure grounding in current knowledge, together with comprehensible access to the latest developments in computer communications and networking.
Emphasis is placed on clear and explanatory styles that support a tutorial approach, so that even the most complex of topics is presented in a lucid and intelligible manner.

More information about this series at http://www.springer.com/series/4198

Florin Pop · Gabriel Neagu
Editors

Big Data Platforms and Applications

Case Studies, Methods, Techniques, and Performance Evaluation

 Springer

Editors
Florin Pop
University Politehnica of Bucharest
Bucharest, Romania

National Institute for Research
and Development in Informatics
Bucharest, Romania

Gabriel Neagu
National Institute for Research
and Development in Informatics
Bucharest, Romania

ISSN 1617-7975   ISSN 2197-8433 (electronic)
Computer Communications and Networks
ISBN 978-3-030-38838-6   ISBN 978-3-030-38836-2 (eBook)
https://doi.org/10.1007/978-3-030-38836-2

This Springer imprint is published by the registered company Springer Nature Switzerland AG
The registered company address is: Gewerbestrasse 11, 6330 Cham, Switzerland

We, the editors and authors of book chapters, dedicate this book to our families and friends with love and gratitude

Preface

Introduction

The value of Big Data applications and their supporting infrastructure like Cloud/Fog/Edge systems lies in the fact that end-users always operate in a specific context: their role, intentions, locations, data handled, and working environment constantly change. According to the research perspective, the Big Data challenges include fundamental research and innovation problems addressing the efficiency, scalability, and responsiveness of analytics services, such as machine learning, language understanding, data mining, visualization, privacy-aware application, *etc.* The existing platforms create an ecosystem based on the convergence of Big Data and Cloud/Edge computing technologies, sometimes combined with HPC for advanced analytics, that in connection with the Internet of Things capabilities enable a wide range of innovations in such sectors as e-learning, healthcare, digitalization, manufacturing, energy, natural resource monitoring, finance and insurance, agri-food, space, and security. In this context, our book, coverage several models and use-cases that are strongly correlated with Big Data challenges.

The book provides, in this sense, an excellent venue for the dissemination of research efforts, analysis, implementation, and final results for Big Data platforms and applications being oriented on case studies, methods, techniques, and performance evaluation, being a flagship driver toward presenting and supporting advance research in the area of Big Data platforms and applications. We are convinced that all authors highlight the results obtained in their research projects and in collaboration with various researchers and practitioners. In the case that the presented work is an extension of already published results, we are more than happy to include the new results in our project.

Book Content

Chapter 1, "Data Center for Smart Cities: Energy and Sustainability Issue", exploiting available dataset recorded in ENEA Data Center (DC), proposes methodologies for energy efficiency evaluation of DCs using appropriate energy and productivity metrics, namely Energy Waste Ratio (EWR) and Data Center Energy Productivity (DCeP). Furthermore, the paper discusses sustainability requirements in the smart city context and evaluate energy productivity at different granularity levels: individual jobs, queues, and DC cluster.

Chapter 2, "Apache Spark for Digitalization, Analysis and Optimization of Discrete Manufacturing Processes", presents digitalization of assembly processes using the latest technologies, analysis of data generated by the monitoring sensors using big data technologies, and optimization of the manufacturing processes by and discuss the research challenges in identifying the steps that have the highest impact on the final output. The main goal is the analysis of the discrete manufacturing processes and more specifically the analysis of the products that are the outcome of the manufacturing processes using as illustrative use case the manufacturing of the regulators in Emerson factory.

Chapter 3, "An Empirical Study on Teleworking among Slovakia's Office-Based Academics", investigates the attitudes and viewpoints of potential teleworkers toward the possibility of introducing teleworking in universities. Moreover, the paper outline some of the key issues related to the implementation of teleworking among office-based academics from a Slovakian perspective. The study makes a significant contribution to a limited collection of empirical research on telecommuting practices at universities and also guides institutions in refining and/or redefining future teleworking strategies or programs.

Chapter 4, "Data and systems heterogeneity: Analysis on data, processing, workload and infrastructure", showcase a survey toward a general understanding of the requirements for handling large volumes of heterogeneous data. Furthermore, it presents an overview of the computing techniques and technologies necessary for analyzing and processing those datasets summarizing the identified key issues for multiple dimensions, including data, processing, workload, and infrastructure.

Chapter 5, "exhiSTORY: Smart self-organizing exhibits", analyzes how the technological advances in the fields of sensors and the Internet of Things can be utilized in order to construct a "smart space". The authors present the system named "exhiSTORY" that aims to provide the appropriate infrastructure to be used in museums and places where exhibitions are held in order to support smart exhibits.

Chapter 6, "*IoT Cloud Security Design Patterns*", evaluates the security issues raised by data-centric elements deployed in IoT networks. The authors present design patterns that are focused on software exclusive solutions for a particular security problem, allowing the use of design patterns on low-end IoT devices, without having to make an assumption regarding the hardware capabilities, like the existence of TPM (Trusted Platform Module) to store and execute the cryptographic operations.

Chapter 7, *"Cloud-based mHealth Streaming IoT Processing"*, presents an overview of architectural approaches and organizational methods to realize a cloud-based mHealth IoT application. While the architecture of mHealth solution using streaming IoT devices is presented along with organizational approaches it copes with extensive data coming with high velocity and volume. Moreover, a use case is presented based on a cloud-based monitoring center that can accept, process, and respond in real time to the demands of real-time monitoring and alerting.

Chapter 8, "A System for Monitoring Water Quality Parameters in Rivers. Challenges and Solutions", identify, and discuss the challenges of implementing a system for monitoring water quality in rivers from continuous data acquisition, to standards compliance and automated pollution detection. Moreover, the authors describe a complete solution for such a system, implemented on Someş River, including data acquisition implemented using WSNs, standard-compliant data storage, data provision services, and automatic assessment of water quality. The proposed architecture is able to support the most important features of a water quality monitoring system.

Chapter 9, "A Survey on Privacy Enhancements for Massively Scalable Storage Systems in Public Cloud Environments", proposes a novel smartphone-based cloud storage encryption overlay, resilient to key theft, trojans, keyloggers, inference, and account compromise. The authors describe the architecture of the system and then other relevant aspects regarding the functionality. The proposed cloud storage overlay will be capable of handling a filesystem-like structure in a manner that will not disclose the actual contents, file sizes or filenames to an adversary.

Chapter 10, "Energy efficiency of Arduino sensors platform based on Mobile-Cloud: a bicycle lights use-case", intends to make a smart device prototype that contributes to efficient use of energy for lights that bikes are equipped. The device determines in real time the degree of agglomeration in the streets and sends data to the cloud for further analysis. This helps analyze traffic on public streets. The prototype has been tested and works very well on the bicycle.

Chapter 11, "Cloud-Enabled Modeling of Sensor Networks in Educational Settings", presents an approach that gives an inner view on conceiving modeling languages with specific applications to sensor networks, supported by configurable tools enabled by cloud. The system is used by students to model the characteristics of sensors and network architecture but also to introduce their extensions through programs that interpret such models. The provisioning process and experimental results for several test scenarios are also described.

Chapter 12, "Methods and Techniques for Automatic Identification System data reduction", describe the Automatic Identification System (AIS) that is utilized in maritime traffic to provide a set of functionalities including, among others, procedures that can help even special occasions including, collision avoidance, and fleet monitoring. The authors present a novel approach for significantly reducing the amount of data produced by AIS without losing the information that could be needed in order to perform real-time data analysis and actions required by it. The proposed algorithm is able. to analyze data and create different kinds of records similar to the video compression algorithms.

Chapter 13, "Machine-to-Machine Model for Water Resource Sharing in Smart Cities", discusses the current initiatives in water management, building an image on what needs are being served, what small or big solutions are being implemented. The model proposed by the authors is a solution for the management of a specific scenario using existing tools which need to be integrated. The second part of the paper contains possibilities of implementation and case studies on the proposed model.

Bucharest, Romania   Florin Pop
March 2021   Gabriel Neagu

Acknowledgments

The editor would like to thank all authors of the book chapters for their valuable contributions and excellent cooperation in the preparation of this project. We express our gratitude and thank them for their hard work.

We address our personal warm regards to Jacek Rak and Anthony Sammes, editor-in-chief of the Springer's Computer Communications and Networks Series, and to Wayne Wheeler, Senior Editor—Computer Science and Simon Rees, Associate Editor—Computer Science, for their editorial assistance and excellent cooperative collaboration in this book project. We add special thanks to Sriram Srinivas for all support and advice, as well as to the editorial and managerial team from Springer for their patience, assistance and collaboration to produce this valuable scientific work.

Finally, we would like to send our warmest gratitude message to our friends and families for their patience, love, and support in the preparation of this volume.

We strongly believe that this book ought to serve as a reference for students, researchers, and industry practitioners interested or currently working in the Big Data domain.

Bucharest, Romania
March 2021

Florin Pop
Gabriel Neagu

Contents

1   **Data Center for Smart Cities: Energy and Sustainability Issue**   1
Anastasiia Grishina, Marta Chinnici, Ah-Lian Kor, Eric Rondeau,
Jean-Philippe Georges, and Davide De Chiara

2   **Apache Spark for Digitalization, Analysis and Optimization
of Discrete Manufacturing Processes**   37
Dorin Moldovan, Ionut Anghel, Tudor Cioara, and Ioan Salomie

3   **An Empirical Study on Teleworking Among Slovakia's
Office-Based Academics**   59
Michal Beno

4   **Data and Systems Heterogeneity: Analysis on Data,
Processing, Workload, and Infrastructure**   77
Roxana-Gabriela Stan, Catalin Negru, Lidia Bajenaru, and Florin Pop

5   **exhiSTORY: Smart Self-organizing Exhibits**   91
Costas Vassilakis, Vassilis Poulopoulos, Angeliki Antoniou,
Manolis Wallace, George Lepouras, and Martin Lopez Nores

6   **IoT Cloud Security Design Patterns**   113
Bogdan-Cosmin Chifor, Ștefan-Ciprian Arseni, and Ion Bica

7   **Cloud-Based mHealth Streaming IoT Processing**   165
Marjan Gusev

8   **A System for Monitoring Water Quality Parameters in Rivers.
Challenges and Solutions**   181
Anca Hangan, Lucia Văcariu, Octavian Creț, Horia Hedeşiu,
and Ciprian Bacoţiu

9   **A Survey on Privacy Enhancements for Massively Scalable
Storage Systems in Public Cloud Environments**   207
Gabriel-Cosmin Apostol, Luminita Borcea, Ciprian Dobre,
Constandinos X. Mavromoustakis, and George Mastorakis

10 **Energy Efficiency of Arduino Sensors Platform Based
    on Mobile-Cloud: A Bicycle Lights Use-Case** 225
    Alin Zamfiroiu

11 **Cloud-Enabled Modeling of Sensor Networks in Educational
    Settings** 237
    Florin Daniel Anton and Anca Daniela Ionita

12 **Methods and Techniques for Automatic Identification System
    Data Reduction** 253
    Claudia Ifrim, Manolis Wallace, Vassilis Poulopoulos,
    and Andriana Mourti

13 **Machine-to-Machine Model for Water Resource Sharing
    in Smart Cities** 271
    Banica Bianca and Catalin Negru

Index 287

About the Editors

Florin Pop (Professor, Ph.D. Habil.) received his Ph.D. in Computer Science at the University Politehnica of Bucharest in 2008 with "Magna cum laude" distinction. His main research interests are in the field of large-scale distributed systems concerning scheduling and resource management (decentralized techniques, re-scheduling), adaptive methods, multi-criteria optimization methods, Grid middleware tools (EGEE, SEE-GRID) and applications development (satellite image processing an environmental data analysis), prediction methods, self-organizing systems, data retrieval and ranking techniques, contextualized services in distributed systems, evaluation using modeling and simulation (MTS2).

He was awarded with two Prizes for Excellence from IBM and Oracle, several Best Paper Award, and one IBM Faculty Award. He is involved in many national projects and international research projects (10+ as project leader). He is an active reviewer for several journals (TPDS, FGCS, ASOC, Soft Computing, Information Sciences, etc.) and he acts as a Guest Editor for several special issues in FGCS and Soft Computing. The results were published in 8 books, more than 20 chapters in edited books, more than 50 articles in major international peer-reviewed journal (15 as main author), and over 100 articles in well-established international conferences and workshops.

He is an active and important member of the Distributed Systems Laboratory (DSLab) in the Computer Science Department. He established and maintains important collaborations with several institutes from EU and around the world: INRIA Rennes-KerData team (France), VU Amsterdam (The Nederland), University Marie and Pierre Curie Paris 6 (France). He is a senior researcher at the National Institute for Research and Development in Informatics, Romania.

Gabriel Neagu, Ph.D., National Institute for Research and Development in Informatics—ICI Bucharest received his Ph.D. in Applied Informatics at the University Politehnica of Bucharest. He graduated a training course for managers of complex information systems at CEPIA (France). Also, he was a visiting researcher on advanced decision support in manufacturing at the Centre for Manufacturing Systems—Institute of Technology, New Jersey, the Robotics Institute—Carnegie Mellon University, Pittsburgh, and the Laboratory for Industrial Process Control—Purdue University, with an IREX (USA) grant support. Since 1995, G. Neagu is senior researcher 1st degree. During this period he has been director of ten competition-based national partnership research projects and beneficiary of two research grants awarded by the Romanian Academy and the Romanian Ministry of Research, respectively. In the same period, he has been a national representative in 13 research projects funded by various European research programs. His list of memberships includes the IFAC 5.1 Technical Committee (since 1996), the National Committee for Research Infrastructures (2007–2011), the European e-Infrastructure Reflection Group (2008–2011), the FP7 International Program Committee for Research Infrastructures, the Research Data Alliance (since 2017), elected member of the Scientific Council of ICI Bucharest (1995–2017), and vice-president of the council (2003–2005 and 2010–2017). He is author/co-author of more than 100 published articles and conference papers. He has been a scientific evaluator for national and EU research programs, IPC member for more than 70 International conferences, invited session organizer and chair at 9 International conferences, and

reviewer for 14 ISI journals. His current research interests include advanced data architectures, data analytics
techniques, e-infrastructures, agile approaches for hierarchical and distributed system development, decision
support systems based on discrete event modeling and
simulation, and open research data management.

Chapter 1
Data Center for Smart Cities: Energy and Sustainability Issue

Anastasiia Grishina, Marta Chinnici, Ah-Lian Kor, Eric Rondeau, Jean-Philippe Georges, and Davide De Chiara

Abstract In a smart city environment, Data Centers (DCs) play a fundamental role, since they enable urban applications by processing big data which comes from inter-connected systems. These processing demands have led to a tremendous increase in DC power consumption. Therefore, the concepts of DC energy efficiency and sustain-ability represent future challenges in smart cities. While assessment of DC energy efficiency with a set of globally recognized metrics is being currently explored, the area of productivity metrics is not thoroughly studied. In particular, there is no general consensus on metrics for direct evaluation of energy used for productive computing operations, or useful work, in a DC. This chapter proposes methodologies for energy efficiency evaluation of DCs using appropriate energy and productivity metrics, namely Energy Waste Ratio (EWR) and Data Center energy Productivity (DCeP) and discusses sustainability requirements in the smart city context. By exploiting the available dataset recorded in ENEA DC, the authors evaluate energy productivity at different granularity levels: individual jobs, queues and DC cluster. Specifically,

A. Grishina · E. Rondeau · J.-P. Georges
Université de Lorraine, CNRS-CRAN, 54000 Vandoeuvre-lès-Nancy, France
e-mail: anastasiia.grishina2@etu.univ-lorraine.fr

E. Rondeau
e-mail: eric.rondeau@univ-lorraine.fr

M. Chinnici (✉)
ENEA, Energy Technologies and Renewable Sources Department-ICT Division, C.R Casaccia
Via Anguillarese 301, 00123 Roma, Italy
e-mail: marta.chinnici@enea.it

A.-L. Kor
School of Computing, Leeds Beckett University, Leeds, UK
e-mail: A.Kor@leedsbeckett.ac.uk

D. De Chiara
ENEA, Energy Technologies and Renewable Sources Department-ICT Division, C.R. Portici
Piazzale Enrico Fermi, 1, 80055 Portici, NA, Italy
e-mail: davide.dechiara@enea.it

© Springer Nature Switzerland AG 2021
F. Pop and G. Neagu (eds.), *Big Data Platforms and Applications*,
Computer Communications and Networks,
https://doi.org/10.1007/978-3-030-38836-2_1

1

portions of energy used for productive computing and energy wasted during computational work are examined. The chapter also provides insights into sustainability of the cluster and proposes a new metric, Carbon Waste Ratio.

Keywords Data center · Energy efficiency · Energy metrics · Energy consumption · ICT · Cluster · Data analysis · Workload management · Smart cities · Big data · Sustainability

1.1 Introduction

A city with pervasive ICT monitoring will be in a process of phasing into a smart city [37, 39]. The recent escalation of big data produced by monitoring systems and smart city applications has contributed to smart city transformation [10, 30, 48]. In the context of a smart city, big data generally refers to large and complex sets of data that represent digital trails of human activities and may be defined in terms of scale or volume, analysis methods and effect on organizations [15]. Cities around the world collect massive quantities of data related to urban living from objects (e.g., IoT), systems (e.g., energy infrastructure) and stakeholders (e.g., residents as energy users). The use of these data contributes to the creation of useful content for various stakeholders, including citizens, visitors, local government and companies. In this scenario, the Data Centers (DCs) play a fundamental role, since they satisfy the processing demands of a vast amount of urban big data which comes from interconnected systems in the cities. However, these processing demands have led to a tremendous increase in energy consumption, and undeniably, electricity usage contributes to the highest portion of expenditure in DCs.

The concepts of DC energy efficiency and sustainability represent future challenges in smart cities, and in the meantime, constitute complex issues in DCs, from the design to utilization stages. Despite the emergence of studies and analysis in the corresponding fields, understanding the energy efficiency and sustainability concerns of DCs as well as their environmental assessment remain limited in practice. Several studies have investigated the use of metrics for DCs in smart cities and identified the relevant set of parameters to assess the energy consumption and evaluate the benefits of energy and sustainability strategies [8, 42, 47]. Nevertheless, a common regulatory framework, which provides standard metrics and methodologies for DCs, is still unavailable [11, 12]. However, some improvement is proposed by authors [17] in terms of a more comprehensive metrics framework and, above all, parameters for direct evaluation of energy used for productive computing operations, or useful work, in a DC [16, 18, 19, 28, 29, 44].

This chapter aims to outline the mutual relation between the energy efficiency strategy and the DC sustainability aspects that together define the requirements for the "smartness" of a city. This current work will encompass discussion on sustainability in the context of both DC and smart city, and the investigation of DC energy productivity on the real example of high-performance computing (HPC) DC cluster.

This research work explores the dependences between smart city, DC and sustainability in the way that smart city depends on DC operations. In the meantime, such dependencies impose certain sustainability requirements on the DC, and in this work, we address DC sustainability in terms of energy efficiency and carbon emissions of the DC. The following objectives will support the chosen aim and elaborate on previous work [28, 29]:

- To provide an overview of DC role in a smart city as an enabler of urban applications;
- To explore energy efficiency and sustainability concerns related to the DCs;
- To discuss sustainability requirements for a DC in a smart city context;
- To propose metrics and methodologies for DC energy efficiency assessment;
- To conduct a case study of a real DC cluster operation that processes smart city applications;
- To choose productivity metrics (e.g., EWR and DCeP) for the DC under consideration and define them in terms of computational productivity of applications independently of the application scope;
- To evaluate energy used for productive computing operations (*useful work*) and *energy wasted* in the real DC at different granularity levels: individual jobs, queues, the cluster;
- To assess carbon emissions generated by DC in the case study.
- To discuss future challenges and opportunities in both energy efficiency and sustainability issues for the DCs in smart cities;

The chapter highlights the importance of DC for processing big data in an urban environment and in the meantime, claims the exigence of a framework based on metrics and methodologies to evaluate DC energy efficiency. This work aims to analyze the real energy consumption of ENEA DC through a set of globally accepted metrics. In particular, the authors aim to investigate the area of energy productivity that is not thoroughly explored, and currently, there is no proposed metric to provide a direct measurement of useful work in a DC. Furthermore, this chapter proposes a methodology that addresses the problem of measurement, calculation and evaluation of energy productivity assessment in a DC, which encompasses both the portion of energy spent on computing processing and *energy wasted* during computational work due to incorrectly processed jobs. It involves the estimation of productive energy consumption by a DC cluster based on the following: statistical data collection and interpretation, software for energy data analysis and mathematical formulation. The current work exploits available data extracted through monitoring of the cluster "CRESCO4" (Computational RESearch Center for COmplex Systems, 4th configuration) in ENEA DC facilities. The dataset covers the power and job schedule characteristics, which have run on the cluster for 1 year. The advancement beyond the state-of-the-art productivity metrics (e.g., useful work) is proposed.

The results of the chapter will help enhance server performance and power management, since appropriate statistical data analysis provides server energy consumption profiles that could be fed into further resource planning. Moreover, the authors evaluate the energy consumed by different queues with several applications. The queues' energy waste has been calculated to provide an assessment of inefficient use of computation-related energy load in the queues with parallel or serial jobs execution. The application of enhanced sustainability metrics with the goal of improving sustainability of a DC is discussed. Additionally, the concept of sustainability in DC operations is investigated through the estimation of its indirect carbon emissions. The authors conclude with recommendations on how the productivity assessment could become the basis of a comprehensive framework to evaluate the energy efficiency of a DC and also proposes consideration for addressing the sustainability challenges. This chapter will contribute to the body of knowledge in establishing the "smartness" of a city from a DC operation point of view, which implies that the processing of urban applications should be energy efficient and sustainable.

This chapter is organized into the following sections: Section 1.2—State of the Art Overview providing an insight into the role of DC in a smart city; Sect. 1.3—Methodology including the description of available real data and methods to achieve objectives from Introduction; Sect. 1.4—Results of the DC Cluster Energy Consumption Analysis; Sect. 1.5—Discussion of the results and sustainability concerns in the DC under investigation; Sect. 1.6—Conclusion.

1.2 State-of-The-Art Overview

The notion of smart cities appears in the twenty-first century with emerging ICT capabilities and rising environmental awareness as a trade-off with improved quality of life. The city is recognized as "smart" if it integrates enhanced technologies in one or several of its components in the following sectors: education, governmental support, healthcare, transportation, safety, clean energy production and other industrial spheres [2]. Solutions deployed in a smart city aim to reduce negative environmental impact and increase the comfort of everyday life. Smart cities' solutions are empowered by technologies that typically rely on interconnected monitoring and reactive components, as well as large quantities of data generated by IoT and other involved systems [3, 4, 30, 37, 40]. Aggregation of historical data and data generated by societal use of applications contributes to the Big Data (BD) phenomenon with characteristics that match at least 3–7 V's versions of a BD definition [26, 33].

Smart cities extensively rely on big data processing thus far primarily provided by cloud technologies, and, therefore, DCs. Characteristics of computational, storage and network resources such as reliability, availability and accessibility of resources, security and optimal power management are crucial for smart cities and their associated applications can impact humans' life and safety. Advancements in fog and edge computing are evidenced [37, 40]. A novel proposal of smart city concept namely *deep urban environment* is introduced by the authors in [20], where a city is based on a new dimension achieved through IoT generating Big Data and Data Analytics enabled by a massively distributed number of sources at the edge. By contrast, this chapter specifically focuses on DCs. It is noteworthy that DCs are also present in fog and edge architectures as a backup resource when edge devices are unavailable. Thus, the link between smart cities and big data is demonstrated in terms of big data processing platform or through edge computing. Limited attention has been accorded to the actual DC operation in this context and a DC is often viewed as a separate area of study. However, this study focuses on DC sustainability, efficiency and optimization of overall DC operations. Additionally, this current work seeks to review the role of DCs in a smart city context. DCs are an integral part of a smart city as an enabler of city services, but at the same time, a huge consumer of energy. Reducing the carbon footprint of DC worldwide is, therefore, a considerable challenge under the pressure of big data deluge [36]. A novel contribution of this work refers to DC sustainability enhancement for better applicability in smart cities through the consideration of DC sustainability issues, monitoring and energy efficiency.

A lot of industrial and research efforts have been dedicated to defining a sustainable DC and, more importantly, to providing suggestions on the incorporation of sustainability goals and practices. The sustainability-related practices and standards encompass Life-Cycle Assessment of DC operations that include equipment, energy and other resources use throughout the DC lifecycle, including its expansion, and upgrade of hardware as well as software components. Several guidelines for sustainable DC operations have been developed by different research and industrial bodies, as well as voluntary programs (e.g., Code of Conduct for Energy Efficiency in Data Centers [1]). They cover renewable energy use, power efficiency in computational and cooling processes, recommendations for appropriate hardware, software, reduced energy consumption and electronic equipment disposal. Specifically, Energy Star program has developed a set of requirements concerning energy use and optimized operations that should be satisfied by IT equipment and its manufacturers to be assigned an eco-label [22]. ASHRAE has developed several guidelines concerning power equipment and DC operational requirements in the pursuit of sustainability[6, 7]. JRC Commission has proposed a holistic framework for assessment of the level of sustainability practices integration in a specific site in its Code of Conduct for Energy Efficiency in Data Centers [1]. The Code of Conduct provides a methodology for DC operators to assess their sites in terms of general policies adoption, IT, power use and cooling efficiency, building exploitation and monitoring. Application of this methodology results in a DC evaluation on the scale from 1 to 5 (best score) in all DC areas that the methodology encompasses. This evaluation also allows DC operators to compare DC performance before and after some sustainability-related actions are undertaken.

Considerable attention regarding DC sustainability is devoted to efficient use of resources, minimization of their wastage and, in particular, energy efficiency Following the specified context, sustainability could be defined as an e-infrastructure strategy that addresses the following [41]:

- Any energy consumption level should be kept as low as possible;
- Any resource should be consumed as effectively and efficiently as possible. In other words, wastage should be minimized;

- Timely and accurate information should be made accessible for the assessment of energy usage, efficiencies and resource use (wastage) to guide and implement process or policy improvement;
- A complete environmental and social impact of activities should be considered;
- The level of IT resource provision should be appropriate to the task being undertaken.

As aforementioned, in recent years, serious effort has been made by consortia involving the industry, academia and public authorities to address the increasing energy demand challenge of the DC sector. Although such effort does provide valuable tools and practices toward reducing energy consumption, they should be merely considered as the beginning of a journey toward environmental targets. In a smart city context, past energy-inefficient practices, such as ignoring the potential use of waste heat or renewable sources, are not sustainable. Now, the research work proposes to plan DC activities according to the forecasted availability of renewable power sources and clean energy from the grid to minimize associated carbon and equivalent emissions [14]. The Real time workload and Delay Tolerant workload developed in [14] could be used with two advantages: (1) better management of task scheduling and (2) better adaptation between DC activities and green energy produced locally (e.g., solar panel on DC roof) for reducing carbon emission.

For the DC sector to continue its operations, energy efficiency and seamless integration within a smart city, appropriate green infrastructures are mandatory steps toward environmental, business and social sustainability. The DCs require adaptation to smart city environment through adoption of energy optimization policies, renewable energy awareness and integration in scheduling process, compliance to both former and new Service Level Agreements regarding the quality of service, quality of user experience and environmental impact [24].

In general, when a DC is not optimized, it contributes to different types of wastage. A DC generates physical waste during refurbishment and upgrade, heat waste as a result of running IT equipment and energy waste due to low computational or cooling productivity in comparison to energy resources used. Moreover, the IT equipment life time may be directly impacted by the temperature inside DC and both energy and cooling management have an incidence about electronic waste produced by DC [50]. Previously mentioned LCA analysis and eco-labeling could be applied to tackle physical waste. Furthermore, thermal energy waste could be reused in the process called heat recovery, when heated water or air in the DC is directed to a heating system (within the DC or nearby buildings) that supplements the existing heating processes [5, 14, 23, 25, 49]. Unfortunately, energy waste caused by the inefficient use of electricity for cooling or computation cannot be reused. Therefore, it is crucial to understand the underlying causes and effects of such waste so as to further minimize it.

The evaluation of DC's impact on the environment, level of different systems' optimization and other characteristics are widely addressed through performance metrics. A growing body of literature has proposed, examined and critiqued the metrics for DC assessment [11–13, 16–19, 21, 28, 29, 38, 44, 45, 47]. For example, in

[47], a taxonomy of the state-of-art DC efficiency metrics is presented for further use by DC practitioners and researchers. A plethora of metrics are categorized (by their DC core dimensions) into groups: energy efficiency (e.g., DCeP, PUE), "greenness" (e.g., CUE, WUE), cooling systems (e.g., HVAC System Effectiveness, Recirculation Index), thermal and air management (e.g., Rack Cooling Index, Return Heat Index, Recirculation Ratio), performance or productivity (e.g., Idle-to-peak Power Ratio, Data Center Performance), security (e.g., Accessibility Surface, Defense Depth), network (e.g., Network Power Usage Effectiveness), storage (e.g., Response Time, throughput) and financial impact (e.g., CapEx, OpEx, ROI) metrics. The authors also discuss the metrics' expressivity and interdependencies in their works. Therefore, this chapter will only cover relevant metrics that are essential for the analysis in this work.

By extending the Sustainability definition (based on the above five guiding principles) to the specifics of sustainable large-scale infrastructure (DCs), it implies that energy consumption ought to be kept at a minimum level as far as possible with available technologies. As mentioned in the Introduction Section, the critical driver of a sustainable DC has embodied within its energy efficiency strategy, which is based on a measurement and control metrics framework. Even though many metrics have been proposed, the debate on a set of globally accepted metrics is still an ongoing challenge, particularly, in the areas of:

- **Productivity metrics**: yet to be explored comprehensively and there is no existing proposed metrics that provide a direct measurement of useful work in a DC;
- **Environmental metrics**: yet to conduct a complete assessment. Waste and Emission metrics facilitate the measurement of the number of natural sources wasted or the quantity of pollution generated by building and managing a DC.

In the latter point, even though there are some carbon and hydro-based metrics such as Carbon Usage Effectiveness (CUE) and Water Usage Effectiveness (WUE) proposed by The Green Grid [8, 42], the deployment of these metrics in a real context is limited. Other groups of metrics emerge from the need to further consider aspects other than the typical simple energy use for energy sustainability. Examples of other elements pertaining to DC sustainability are: efficiency of a single sub-system, the existence of onsite renewable energy sources (RES), recovery of energy wastes and end-of-life e-waste handling (waste production). The metrics related to these aspects are summarized in Tables 5 and 6 in [17].

1.3 Methodology

In this work, the authors employ real-use case analysis and action research to develop a methodology for DC energy efficiency assessment. To achieve this aim, productivity metrics are investigated with consideration of available disaggregated data. This set of data are related to the power consumption of operating systems of HPC Cluster CRESCO4 hosted by DC within the ENEA Portici Research Center. In this work, power measurement analysis and related policies have been further improved with respect to the previous study conducted on data available for CRESCO4 [16, 18, 19, 28, 29].

The authors apply a quantitative technique to evaluate the cluster energy use based on available data on jobs scheduling and power consumption. Furthermore, the authors discuss what part of jobs processing is assumed as *useful work* (or *work done*) within the scope of the current analysis. The definition of *work done* will further help identify proportions in which energy is consumed for *useful work* and wasted for jobs, which have not brought about useful results. In particular, to assess energy spent on the *useful work*, the authors apply Data Center energy Productivity (DCeP) metric in a way that is compatible with the cluster operations and available measurements. Also, the authors focus on waste energy estimation in DC and repartition waste energy according to the types of jobs that have caused this waste. The analysis continues with a lower level of granularity, where the calculated energy consumption is associated with individual applications running on the cluster. This association facilitates the estimation of energy performance characteristics of queues as well as parallel and serial jobs. Finally, the energy consumption profiles are interpreted from a sustainability point of view.

1.3.1 Data Center Facilities and Dataset Description

The assessment of energy consumption, the calculation of *energy waste* and the queue analysis are based on available data gathered on CRESCO4 cluster of ENEA DC during the period from February 2017 to January 2018. The cluster CRESCO4 consists of 38 Supermicro F617R3-FT chassis, with eight dual CPU nodes each. Each CPU is of the type Intel E5-2670 and hosts in its turn eight cores, which results in a total number of 4864 cores. The CPUs operate at a clock frequency of 2.6 GHz. Furthermore, each core of the system is provided with a RAM memory of 4 GB. Computing nodes access a DDN storage system, constituting a total storage amount of 1 Pbyte. Computing nodes are interconnected via an Infiniband 4xQDR QLogic/Intel12800-180 switch (432 ports, 40 Gbps).

More specifically, the analysis correlates accounting data from two available datasets: Platform LSF (Load Sharing Facility) job scheduler and the corresponding power consumption collected from installed PDUs obtained through Zabbix monitoring tool. Briefly, LSF is a workload management platform and job scheduler, for distributed HPC systems. This platform is concerned with deciding which process is to be run and is designed to keep CPUs as busy as possible. It stores a log file that contains information on executed jobs and the usage of computing nodes (cores). The information comprising the LSF dataset includes the number of cores and queue name assigned to every process (to clarify, the words process, job and application are used as synonyms in this work), start and stop time of the application processing, names of executable file and directory and the marker of the process final status: "done" when it ends successfully or "exit" when ending with an error. The jobs are

Table 1.1   Size and period covered by datasets

Dataset	From, DD-MM-YYYY HH:MM	To, DD-MM-YYYY HH:MM	Number of samples
Before preprocessing			
LSF	11.11.2016 17:38	25.01.2018 04:12	529,512
Zabbix	19.02.2017 12:00	19.02.2018 14:00	8763
After preprocessing			
LSF	19.02.2017 12:00	25.01.2018 04:12	519,095
Zabbix	19.02.2017 12:00	25.01.2018 04:00	8153

also divided by their mode of execution: they are recognized as serial ones if they only use one core, and parallel, otherwise.

Our experiment uses the LSF log file to understand in what way the cores are occupied by users and the Zabbix database to correlate the processing load with power consumption profiles during job execution. In detail, the Zabbix dataset contains average level of power consumption, minimum and maximum registered power consumption for each hour between ranges indicated in Table 1.1. The LSF dataset covers the details about the number of cores assigned by the scheduler for every process, start and end time of the application activity, names of executable file and directory and the marker of whether the process has finished successfully ("done") or with an error ("exit"). The second dataset covers IT jobs running within the interval reported in Table 1.1. From these datasets, the time intersection is taken and covers 11 months from 12:00, 19th of February 2017 to 12:00, 25th of January 2018, divided by 19th day, 12:00, of each consecutive month except January 2018. The resulting time intervals of dataset coverage are reported in Table 1.1, where the number of rows (samples) diminished after applying the intersection.

The task scheduling of the cluster is based on First Come First Served algorithm—a basic scheduling policy in which tasks are served in the order of their arrival in the system. This strategy reduces the waiting time for tasks. The cluster's scheduler allocates jobs to 18 different queues, 11 of which accommodate only parallelized jobs, 3 other queues are designed exclusively for serial jobs, and the remaining—for both parallel and serial jobs. Around 92% of all submitted jobs are processed in a serial mode, which has left room for only 8% of jobs being calculated with parallelization techniques. The queue characteristics are reported in Table I in [29].

In all the queues, approximately 40 types of applications that are running on the CRESCO4 cover several fields of research, such as climate research, renewable energies, environmental issues, materials science, efficient combustion, nuclear technology, plasma physics, biotechnology, aerospace, complex systems physics, HPC technology. Moreover, many other kinds of applications are embedded in scripts that, through libraries, recall the functionality of consolidated software suites.

The following subsection explains how the two datasets are used for the energy consumption estimations. Data analysis and computations required for the purposes of the current work are performed in Python programming language, suitable for big datasets.

1.3.2 Data Analysis

To achieve the goal of the work and estimate effective energy consumption by the cluster and its waste, the energy conservation law has been expressed in terms of available characteristics of the DC cluster:

$$\sum_{i=1}^{K} \int_{t_{i,j}^0}^{t_{i,j}^1} c_{i,j} \cdot x_j dt = E_j, \quad j = 1, \dots, N. \tag{1.1}$$

Here, x_j is the main set of variables expressing the power required from one core to process an arbitrary application every second during the hour j. This set of variables is an unknown target for energy consumption estimation. Its multiplication by $c_{i,j}$, the number of cores required to work on application i during the hour j, gives the power required by the application i during the same hour. So, the ID of a process is denoted by i, and K stands for the number of all processes active during the considered month. Then, integration over time with the limits $t_{i,j}^0$ and $t_{i,j}^1$—start and end time of the application i activity during the hour j—results in the application energy demand, which is further summed over all the applications processed during the hour in question. On the other hand, the cluster consumes E_j watt-hours of energy during the hour j; and N stands for the number of hours in the extracted month. This equation is then rewritten to avoid integration over non-continuous variable:

$$\left(\sum_{i=1}^{K} c_{i,j} \cdot \frac{t_{i,j}}{3600} \right) \cdot x_j = E_j, \quad j = 1, \dots, N, \tag{1.2}$$

where the value $t_{i,j} = \left(t_{i,j}^1 - t_{i,j}^0 \right)$ describes the duration of process i activity expressed in seconds. This explains the need to divide it by the number of seconds in one hour for consistency of dimensions. Thus, the interpretation of the energy conservation law is a system of linear equations with one unknown each. The equations can be resolved by simple division of the right part by the sum from the left part to obtain the solution for $x = \{x_j, j = 1, \dots, N\}$.

For the next step of data analysis, the obtained solution is cleansed of outliers, which possibly appear as a result of an incomplete set of measurements. To explain, the quantiles are applied as in the following formula to identify the outliers:

$$|x_j - \bar{x}| > 2 \cdot (Q_3(x) - Q_1(x)), \tag{1.3}$$

where $\bar{x}$ denotes the mean of the values of the vector $x = \{x_j, j = 1, \ldots, N\}$. $Q_1(x)$, $Q_3(x)$ stand for third and the first quartiles of the vector x correspondingly, under the assumption of x being normally distributed, i.e. the values which separate 25% of lower values of x and its highest 25%. The values $Q_3(x)$ and $Q_1(x)$ separate a set of x values into four subsets of equal size, since quartiles are defined as the fourth quantiles. To this end, Euclidian distance of x_j from the mean value $\bar{x}$ is calculated and compared with double distance between two values of the two densest subsets, on which quantiles divided the values of vector x. The multiple of 2 is chosen empirically, since there is no fixed algorithm to find outliers in every problem. To clarify, the exclusion of data is not the priority here, because the more data are excluded the less accurate is the energy consumption estimation.

Data inconsistency has been further studied so that the sum on the left-hand side of the Eq. 1.2 does not equal zero, when the right-hand side is not null either. In such cases when the sum $\left(\sum_{i=1}^{K} c_{i,j} \cdot t_{i,j} \right)$ is equal to zero, it has been assumed that the data reported by LSF during the hour j are insufficient and have not been considered for the final output. The same situation could hypothetically have occurred during the hour when no job was submitted to any queue and thus designated the idle power. This hypothesis is checked and rejected because of the following findings. Average energy consumption of the periods when no processes are registered to have been active is 42 kWh, i.e. when the aforementioned sum equals zero, and changes to 47 kWh when those hours and corresponding data are excluded from the dataset. Moreover, the range of energy consumed, when no process has been active, lays within the range of energy consumption when cores have been reported to work on the jobs. Namely, if all hours with no processes registered to be active are united into one dataset, the range of energy consumption is [27.3; 58.8] kWh, whereas in the dataset with non-zero sums of coefficients for each hour, the energy consumption lies within the interval of [14.4; 65.5] kWh. This inclusion does not allow any estimation on idle energy consumption, because it lies within the energy consumption range reported for the hours with active processes running.

The system in Eq. 1.2 is chosen after consideration of its more granular analog:

$$\sum_{i=1}^{K} \left(c_{i,j} \cdot \frac{t_{i,j}}{3600} \cdot x_{i,j} \right) = E_j, \quad j = 1, \ldots, N, \tag{1.4}$$

the system in Eq. 1.4 differs from the chosen system in Eq. 1.2 only in the aspect that unknown power consumption per core each second of any arbitrary process $x_{i,j}$ is associated with an individual process i in the system of Eq. 1.4 and denoted as $x_{i,j}$. The expressions in Eq. 1.4 become SLAE of $N \times K$ dimension, where as previously K stands for the number of applications with registered activity during hour j, $j = 1, \ldots, N$, contrary to the previous system of Eq. 1.2, when the dimension of the solution x_j was $N \times 1$. The system of Eq. 1.4 with increased granularity was expected to provide more precise results, nevertheless, the study of such system disclosed some characteristics, which did not allow to obtain a relevant solution.

Consider the system for the month of 19th February–19th March 2017 as an example to investigate the system of Eq. 1.4 characteristics preventing its use. This system has a sparse matrix with non-zero elements comprising only 3% of all values. The conditional number of the matrix, obtained through the SVD-analysis, is of the 10^{16} order. Additionally, the variables $x_{i,j}$ are constrained to be non-negative, because it is assumed that nodes do not produce electrical energy. These SLAE properties and constraints have resulted in poor accuracy of the solution, although it was obtained with the algorithms designed specifically for ill-posed problems, i.e. problems with a large conditional number of the matrix. The tested algorithms enumerate Least Squares, Least Squares with regularization in L^2, Least Squares with regularization in L^2 and L^1 algorithms realized in Python Scikit-learn library, linear_model module. With this in mind, a decision has been made to omit the index i for individual processes differentiation, thus decrease the dimension of the system and generalize the unknown variable x_j to the power consumption for all jobs registered for the hour j. The vector x_j is still useful for calculating energy consumption of individual processes by multiplying it and corresponding weights $c_{i,j}:(c_{i,j} \cdot t_{i,j}/3600)$.

The mathematical formulation and solution of the system of Eq. 1.2 are further used for DC energy metrics evaluation, since the solution provides energy consumption of every application during every hour per month and might be accumulated to the desired intervals of 1 month or the general period with available data. Owing to the fact that the data also contain flags of how successfully each job is finished, it brings additional values to assess the energy spent on *useful work* and wasted for incorrectly finished jobs or their parts. The assessment will be further discussed in the following parts of the chapter. All the energy consumption inferences about the DC cluster aim to increase awareness of the DC operators about energy profiles and distribution of *energy waste* between submitted jobs and in time, so as to enable suitable actions for the DC improvement.

Further analysis comprises the following steps:

1. Calculate:
   - values x_j for each month to find the power consumption of any arbitrary process during 1 second from the Eq. 1.2;
   - $(c_{i,j} \cdot x_j)$ and sum these values for all the hours j when the process i was active. On this step, energy consumption of every process is obtained within a month.

2. Make a distinction between successfully completed jobs and those ended with a type of error, which will be further associated with useful work and energy waste markers.

3. Merge jobs by queues to evaluate the proportions of their submissions in different queues, apply DCeP and EWR metrics to estimate the efficiency of energy consumption on the queue level of granularity.

4. Use data on parallel/serial modes of jobs execution to evaluate proportions in which users submitted such jobs, determine their energy consumption and EWR metric values associated with two modes of execution.

5. Moving to the higher level of the whole cluster, estimate energy consumption
    of jobs which produced useful work and energy waste:

    – Obtain values for monthly energy consumption of these two groups of jobs;
    – Translate energy consumption into approximate amount of CO_2 emissions
       per month using a carbon factor for Italy and categorize the emissions by the
       purposes, to *useful work* and energy waste.

The next subsections are devoted to the metrics applied in the data analysis and
the definition of *useful work* and *energy waste*.

1.3.3 Metrics

Monitoring of energy usage and consumption is essential to pursue the energy efficiency target and, in the meantime, to reduce the *energy waste* in the DC operations.
For this reason, metrics are required to provide insight into how efficient energy is
using in the computing operations in a DC. Consequently, productivity metrics (e.g.,
DCeP) have been created to address this challenge. This category includes indices
related to the quantity of the *useful work* within a DC from an IT perspective. Hence,
these indices should lead to questions such as: What is the *useful work* of a DC? and
How does one calculate the *useful work* of a DC [17]? Even though many metrics
have been proposed, the debate on a set of globally accepted metrics included the
productivity metrics is still an ongoing challenge. To address this issue, the authors
explored the productivity metrics area and provide a direct measurement of *useful
work* in DC in this work. In detail, the cluster's energy consumption is evaluated
through the DC energy productivity (DCeP) metric [21, 45], which is expressed as
follows:

$$DCeP = \frac{\text{Useful Work Produced}}{\text{Total DC Energy Consumed over Time}}. \tag{1.5}$$

In our experimental setting, the available data for *Useful Work Produced* refer to
energy spent on correctly completed jobs processing. Total DC Energy Consumed
over Time is represented here by the energy used for all jobs (both prematurely
aborted and correctly completed ones). Although the generally accepted practice is
to consider energy, which goes for both cooling and IT systems, under the notion
of Total DC Energy Consumed over Time, limitations of the data retrieved from the
cluster do not provide sufficient information for such study.

The estimation of the *energy waste* (or *no work*) in DC is supported by the Energy
Waste Ratio (EWR) metric [16, 38] that is expressed as follows:

$$EWR = \frac{\text{Energy Wasted for Not Useful Work}}{\text{Total DC Energy Consumed over Time}}. \tag{1.6}$$

This metric assesses the ratio of energy spent on the work, which has not provided any useful result, or on the *not jobs* from the *energy waste* categories, which will be covered in the next paragraph.

1.3.4 Energy Waste Analysis

The analysis of the *energy waste* provides insight into ways to reduce the overall energy consumption in DC and subsequently improves its power management by additionally employing workload analysis. To achieve this aim, the work also focuses on categorizing submitted jobs to distinguish between the jobs, which result in productive work done, or *useful work*, and the jobs (or the *not useful jobs*) that represent only inefficient work with their inefficient energy use, or energy waste. The latter jobs can be subdivided into three categories to assess their contribution to wasted energy:

- Jobs with maximum running time of 30 s (category I);
- Jobs that exceeded the queue time (category II);
- Jobs that quitted the queue with an error for any other reason (category III).

Category I comprises jobs with such short running time, which occur to represent the work of the scheduler only, while the scheduled application itself has not been started. The value of the threshold, set at 30 s, is an empirical choice based on the knowledge of the pre-working time of the LSF application and then its dataset. The jobs running during less than 30 s represent the preprocessing phase, and they have not produced any *useful work* in terms of results for the end-user, who had submitted them to the cluster. For this reason, these jobs are considered to cause *energy waste*. Given that the running time of the majority of jobs has varied from 2 seconds to 221 h, the average is 2 h, the processes from the category I have consumed small amounts of energy. However, the presence of such processes still affects the cluster work.

Category II consolidates all jobs the running time of which exceeds the maximum queue time. The existing policy of the DC usage states that if a job is submitted to processing units and allocated into a specific queue by LSF, the queue allows this job to run for a particular time. In case of exceeding the maximum time assigned by the queue, the job is removed from the queue, being reported as an erroneous process, sometime after the queue, time limit is exhausted. However, while the job is being processed within the queue time, it produces results and cannot be regarded as a reason for *energy waste*. For example, consider a job with an exit status that was running on the queue allowing the maximum period of 600 s. If it started at 1,494,316,196 Unix time and ended at 1,494,317,094 Unix time on this kind of queue, the total job duration (reported stop time—start time) is 898 s and exceeded the max queue time. Hence, in our analysis, we calculate as *useful work* the work associated with the part of the job that was running for total queue duration (600 s); meanwhile, we consider as *energy waste* the part of energy spent on the job running in the rest of the time 898–600 = 298 s. The energy is wasted only when the job runs after the queue time limit, which is the focus of category II. Then, category III is composed of jobs with any other malfunction causing jobs interruption both by end-users or by the system.

These three defined categories are further used to measure Energy Waste Ratio (EWR) metric (Eq. 1.5) where, the term *Not Useful Work* or *energy waste* refers to the jobs that ended with an error for one of the aforementioned reasons.

1.4 Results: DC Cluster Energy Consumption

The methodology proposed in the previous sections has been applied to the data acquired for the CRESCO4 cluster of ENEA DC. The results obtained through data analysis cover energy consumption of individual jobs, queues groups of jobs divided by the mode of execution and the whole cluster. The added value of this work is the contribution made toward DC energy efficiency assessment that comprises different levels of granularity for the estimation of energy consumption (i.e. starting from individual jobs and finishing by the whole cluster). Second, it is noteworthy that a portion of IT energy is not provided for *useful work*, although it is spent on IT equipment operation. Categorization of jobs causing *energy waste* and assessment of their contribution in energy use is aimed at raising awareness of the DC operators for IT energy consumption. The categorization is also devoted to inspiring more precise analysis of DC operations, particularly, during DC energy efficiency assessment, which should identify possible weak points in order to trigger improvements. Visual support for these inferences and exact values for the DC cluster in question are provided in the subsections below.

1.4.1 Energy Use by Applications

Workload distribution between tasks occupying the reported DC cluster is illustrated in Fig. 1.1. In detail, the figure shows the proportion in which energy is consumed by different processes over the overall period of monitoring. In the meantime, it indicates the purposes of cluster computations. As observed in Fig. 1.1, the cluster processes tasks that are not exclusively dedicated to smart cities. For example, air quality monitoring and climate modeling share the cluster with Monte Carlo algorithms for simulation in particle physics. Moreover, some initial versions of smart home and other urban applications are being developed and only consume negligible amounts <1% of the total energy not indicated in the figure. This variety of applications is typical for a data center which is adapted for smart city purposes. Many DCs have already operated for different purposes, so it might be efficient in terms of cost and resources that could otherwise be used for installation and construction to reuse available computational, storage and network capabilities of existing DCs.

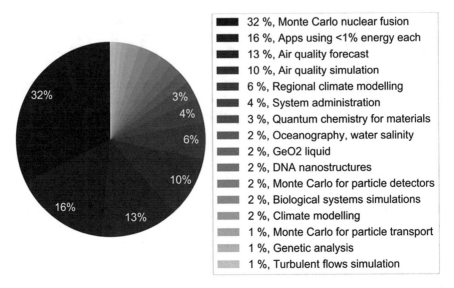

Fig. 1.1 Ratio of energy used by applications measured during the overall period of 11 months. Applications are grouped according to their intended scope. Ratio is calculated with respect to the energy used by all applications within the overall period in question

Out of all processes, the statistical Monte Carlo methods for particle detection, transport and nuclear fusion are registered to consume the most energy, which is reasonable, considering that they require a large amount of random number generations. Air quality simulation and forecast together are responsible for 23% of the cluster energy consumption, while the previous category portion is 35%. These two groups form up to more than half of the cluster total consumption of energy resources, which define the main research orientation of the cluster processing. Other applications individually do not require greater than 6% of cluster resource use, while the smallest considerable portion of energy is dedicated to genetic analysis and mathematical algorithm for turbulent flows simulation. The existing applications that require <1% of energy consumption over the period in question (11 months) have been combined into one group. Given that this group necessitates 16% of total cluster energy, the cluster utilization pattern is visible. It is probable that the cluster processes a large number of applications with low energy demands.

1.4.2   Energy Analysis of Queues of Jobs

As previously mentioned, all the jobs are distributed into queues, according to their order of submission, parallel or serial mode of execution and FCFS policy. Therefore, on the next step of data analysis, it is proposed to uncover mutual relations (if any)

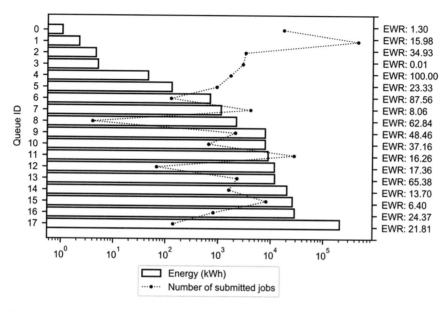

Fig. 1.2 Energy consumption (kWh) of queues estimated for the total period of 11 months, EWR metric is indicated in % of total energy consumption of each queue, number of jobs submitted to each queue

among energy use, the number of submitted jobs and *energy waste* ratio for individual queues.

The energy load for queues, number of jobs submissions to the queues and EWR in percent are represented in Fig. 1.2 with bars, dotted line and notes on the right side of the graph. Energy consumption of queues ranges from 1 kWh to 207 MWh over the total period of monitoring. However, there is no clear correlation between energy consumption of queues and EWR values estimated for them. As a clarification, the values of EWR metric have been calculated for every queue separately, i.e. queue EWR is the ratio of energy consumed by jobs defined as *energy waste* cause in the methodology section over the total amount of energy used by the queue for the overall period under consideration without division on months. One negative factor regarding the queue EWR estimation is that one queue is reported to have 100% value of this metric. However, EWR average over the queues remains at 32.5%. These values might be useful for the comparison of the cluster's work, but it poses a challenge to estimate how small or big the standalone value is for a specific data center.

Further analysis considers the number of job submissions into the queues also depicted in Fig. 1.2. Owing to the fact that one job may have been submitted several times to one or more queues, the word "submission" is used here to cover all actions of submission. For example, if one application is submitted *n* times by the same user at different moments, the number of total submissions is increased by *n*. It is also worth noting that users do not choose the queue for submitting the jobs, instead, all submissions, or rather allocations of jobs to queues are conducted by LSF jobs scheduler. Job allocation and the estimation of operational efficiency will be a goal of this part of the analysis. As a result, the queue number 17 with the second smallest energy consumption over the total period and EWR of 16% is reported to have had the most significant number of job allocations. Another finding is that 99% of energy is consumed by 9.5% separate submissions. These values are obtained by grouping consumption and submission values of the first 14 out of 18 queues.

Essentially, observations of high EWR values could be employed for enhanced notification and scheduling systems, which would inform the user of repetitive malfunctioning jobs. A typical user behavior in case of a job failure is to terminate and resubmit it. However, this incurs ineffective use of computer resources and increases the probability of *energy waste*. A good practice is to test and debug programs and guarantee that they work properly and produce the expected results. Additionally, it is crucial to understand when it is better or worse to implement a serial or parallel job. When a parallel execution modality is chosen, it is recommended to optimize the algorithms for the use of all available resources, which brings the analysis to its logical continuation with the investigation of parallel and serial jobs execution.

1.4.3   Energy Use by Parallel and Serial Jobs

Consideration of the two categories of execution modes, namely parallel and serial mode, fosters a higher granularity analysis of job characteristics. Therefore, the focus is moved from the queues to these two categories of jobs and covers data for every month. Analogously to the previous analysis, parallel and serial job groups are explored in terms of energy consumption, number of submissions and EWR as shown in Fig. 1.3. This figure is a result of the integration of three separate graphs of the same job characteristics in Figs. 4, 5 and 6 of [29], combined with the intention to provide a better visual support for comparison of jobs' activity throughout the months in question. Thus, in the current work, the vertical axis of Fig. 1.3 is associated with 11 months: each month covers data from the date 19th of the first indicated month name to 19th of the second-month name, for example, the tick "Feb-Mar" corresponds to the period from the 19th of February 2017, to 19th of March 2017. Horizontal bars present energy consumption in kWh by parallel (gray bar) and serial (dotted bar) jobs. A number of submissions of parallel and serial jobs are represented with a solid and dotted line correspondingly. Finally, the EWR values are associated with each month and mode of jobs' execution and are expressed in percent.

The energy consumption and EWR of parallel jobs generally prevail over serial jobs, while the number of serial jobs submissions is observed to have been higher than parallel jobs submissions in 10 out of 11 months. An even pattern of parallel jobs EWR is depicted in Fig. 1.3 with a mean value of 22%, whereas the same metric for serial jobs fluctuates between 0.025 and 4%.

In essence, Fig. 1.3 depicts a general trend for energy consumption. It is noted that the monitored cluster parallel jobs consume more energy and, if such a job fails,

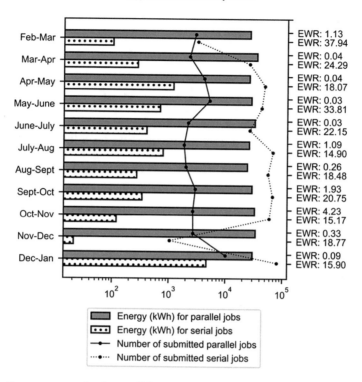

Fig. 1.3 Energy consumption by parallel (gray) and serial (dotted) jobs, number of submitted parallel (solid line) and serial (dotted) jobs, and EWR values for each group of jobs reported for different months

then the energy required for computations until to the failure point is largely wasted in comparison with serial jobs. In addition, serial jobs consume around two times less energy throughout the studied period, although they are submitted 200 times more frequently on average, the values have been dispersed throughout the months from 10 to 1000 times.

The EWR values shown in Fig. 1.3 are less than 40%, which indicates that more than the half of energy has been used for producing *useful work*. However, this relation does not allow to examine how effectively the energy has been used within the execution mode groups. For this purpose, it is suggested to apply modified EWR formula for our particular case:

$$\text{Modified EWR} = \frac{\text{Energy Wasted by the Type of Jobs}^*}{\text{Total Energy Consumed by the Type of Jobs}^*}, \quad (1.7)$$

*where $Type\,of\,Jobs$ is $Parallel$ or$Serial$. To draw a distinct comparison, the initial EWR values shown in Fig. 1.3 have been calculated with the relation to overall monthly energy consumption of parallel and serial jobs comprising both useful work and *energy waste*. By contrast, Eq. 1.7 for the modified EWR *energy waste* by a parallel/serial group of jobs is normalized by energy consumed exclusively by the same group.

The values of modified EWR are calculated for two categories in question for every month and are represented in Table 1.2. The *energy waste* within the category of parallel jobs remains stable around 20% for the majority of time with two peaks in September–October and December–January. This trend is dissimilar to the one of the modified EWRs for serial jobs, as the values of the latter fluctuate dramatically throughout the period. These unsteady values add to the unpredictability of the serial job processing: they are submitted in larger numbers than parallel jobs but contribute (from low 2% to a peak of 97%) to the serial jobs group's *energy waste*.

1.4.4 Assessment of Useful Work

As mentioned in the previous sections, the DC energy consumption has increased dramatically over the last decade, and this situation has determined the quest for metrics that evaluate DC energy efficiency. Despite a great interest, traditional metrics for measuring energy efficiency in DC (e.g., PUE) are limited to calculating the energy required for the major IT components of the DC plus the energy for supporting infrastructure. In contrast, the present section aims to compute energy efficiency metrics based on a clear definition of the *useful work* which is a parameter intended to gauge the real computing carried out by a DC.

The analysis on productivity metrics related to the *useful work* is also necessary to achieve sustainability goals. Generally, *useful work* of a DC is represented by overall computing activity of the IT Equipment (ITE). The ITE activity comprises computing, storing and transferring data and is referred as IT services. Appropriate productivity metrics are used to measure and assess such activity's characteristics [17, 45]. Nevertheless, productivity metrics differ in their approach to assess *useful work*. As a consequence, none of the metrics has provided a practical way to exactly calculate the work done or *useful work*, even though several attempts have been made to define the productivity metrics for DCs. Among all the productivity metrics, DCeP is the most significant one [47]. The present stage of work is devoted to calculating it based on the DC operation data.

DCeP metric evaluation is facilitated by the consideration of each IT job power consumption per core during each second obtained from the Eq. 1.2 and the information about the fulfillment status of jobs. Thus, monthly energy consumption with separation on the energy for *useful work* and *energy waste* is obtained for each month in the 11-month duration. Furthermore, DCeP is evaluated as the ratio of energy for *useful work* over the total energy consumption. The results are represented in Fig. 1.4. The largest portion of energy use is observed during the month from 19th of March to 19th of April reaching the point of 35.6 MWh, whereas the smallest portion of energy consumption is reported in the months from 19th of July to 19th of September. DCeP

Table 1.2 Modified energy waste ratio for parallel and serial jobs

	Feb-Mar	Mar-Apr	Apr-May	May–June	June-July	July-Aug	Aug-Sept	Sept-Oct	Oct-Nov	Nov-Dec	Dec-Jan
Parallel jobs	16	18,9	15,9	21,3	18,7	15,3	22,4	34,2	18,1	24,3	43,8
Serial jobs	24,4	41,6	97,4	80,8	21,2	37,2	2,7	2,2	11,3	64,7	8,5

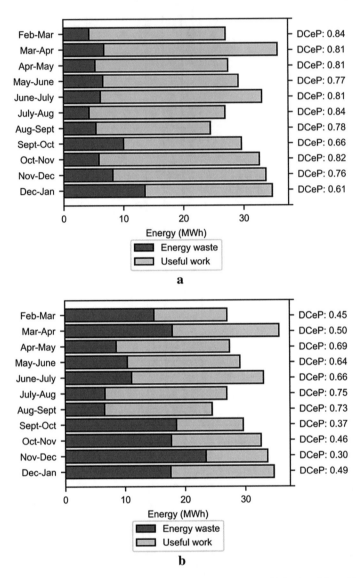

Fig. 1.4 Monthly analysis of energy consumed by correctly finished jobs (*useful work*) and jobs, which exited a queue with an error status (*energy waste*), and DCeP; **a** Energy *waste* categories are considered, **b** Jobs are not categorized by causes of *energy waste*, data on jobs status are taken directly from LSF

Fig. 1.5 Ratio of executed jobs that consumed less than 10 kWh per month and ended either with or without an error in correspondence to the overall number of jobs

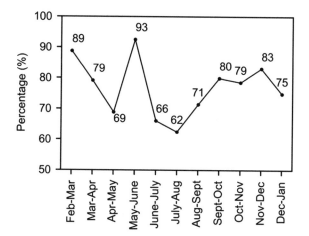

varies from the minimum of 0.61 in the last reported month to 0.84 in the June–July period.

As a note on data analysis strategy, in the case when jobs are taken directly from the LSF data, without categorization and identification of additional categories I and II from subsection Energy Waste Analysis, DCeP is reported to stay at a lower level than after preprocessing the LSF dataset and extracting categories. DCeP differences can be observed in Fig. 1.4b, where no categorization has been done, versus Fig. 1.4a depicting values when categorization has been considered. These results are explained by the fact that, in the raw LSF dataset, all the jobs exceeding the queue time are marked with "exit" status and referred to *energy waste*. However, as described previously, the energy used within the queue time had been spent on *useful work* and only the remaining part of the processing period caused *energy waste*. Also, some jobs with duration time within 30 s were marked as *useful work*, which, according to our assumptions, is not true. Henceforward, the categorized dataset is used, i.e. the one corresponding to Fig. 1.4a.

Energy consumption of the processes is found to have been unevenly distributed. The majority of the processes consumes less than 100 kWh per month. A more granular analysis showed that from 62 to 93% of the overall number of the cluster jobs consume less than 10 kWh per month as shown in Fig. 1.5.

1.4.5 Assessment of Energy Waste

Energy waste assessment has been addressed in academia and industry both qualitatively and quantitatively, for the reason that inefficient energy use causes increased electricity cost and negative environmental impact if the extra energy used is produced from non-renewable resources. Some research work explores VM allocation-related *energy waste* that is particularly crucial for cloud paradigm in DCs, which provide computing resources to users in the forms of infrastructure, platform and software as a service. Such work proposes Virtual Machine (VM) allocation strategies and algorithms, which increase the performance and QoS characteristics of DCs [9, 34, 46]. In other research work, *energy waste* is discussed in terms of heat generation, and in such cases, thermal energy reuse is suggested as a potential solution. For example, the heat recovery in smart cities can be used for heating (sometimes partially) the nearby buildings, or even the premises of the same DC to provide good working conditions for offices within DC premises [23, 25, 49]. Furthermore, in the research that approaches *energy waste* with the use of metrics similarly to this work, useful energy, as the opposite to the *energy waste,* might have an ambiguous definition. *Useful work* is identified on the application level, varying from the number of floating-point operations, number of service invocations, number of transactions, or another essence related to the individual application [21, 47]. In [27] the authors classify tasks failures-based causes such as server or software failure, scheduler issue and evaluates energy spent on such tasks, but does not explicitly use EWR metric, which has emerged later.

The present work aims to define *useful work* and *energy waste* of computing activities and in particular for all the range of applications, which are processed by the target cluster in a unified manner and highlights the importance of metrics used for quantitative evaluation. Additional outstanding point of the current work is that the real data from a real DC are used for analysis, thus, in comparison with simulations of a DC operation, it shows real issues, which should be addressed both by DC operators (to manage the processes and data acquisition), and by the user side to improve their applications.

In this part of the work, we investigate the *Not Useful Work* or *energy waste* to calculate how much energy has been used for computing activities but has not produced *useful work*. *Energy waste* and energy spent on useful computational work are two supplementary portions together forming the total cluster ITE energy consumption. For this reason, EWR is studied for individual applications and *energy waste* categories, while monthly EWR values are not shown to avoid repetition, since they are equal to $(1 - DCeP)$. In addition, job distribution into waste energy categories is analyzed both from their energy consumption perspective and their share in the number of submitted jobs. Statistical characteristics are taken from the monthly samples of data and are shown in Table 1.3. The table includes the minimum, maximum, mean value and standard deviation of the ratios of energy used by jobs from each category related to the general energy use. As might be observed from Table 1.3, processes with short running time consume the least share of energy (i.e. approximately 0.03%), whereas jobs that exceed the queue time used around 0.2% of total energy consumption. Both categories have a small deviation from the mean value, which signifies moderate fluctuation of their energy use over the months. In contrast, processes malfunctioning for other reasons (category III) used a range of 16–39% of energy. The deviation of the latter category is the highest, which highlights the necessity of closer investigation into incorrect processing of jobs to decrease the number of their submissions and increase energy productivity of the cluster. Values of EWR for all the three categories are associated with applications and represented

Table 1.3 Energy waste ratio by job categories with relation to overall energy, %

Statistical characteristics	(a) Running time ≤31 s	(b) Running time > queue time	(c) Other reasons
Minimum	0.007	0.004	16
Maximum	0.06	0.3	39
Mean	0.03	0.2	23
Standard Deviation	0.01	0.09	7

in Fig. 1.6 in %. When compared with Fig. 1.1, the pattern is similar: Monte Carlo analysis has consumed most of the energy over 11 months causing the most significant *energy waste* of all applications. Applications that require small amounts of energy correspondingly create low energy loss, for example, genetic analysis and turbulent flows simulations.

Monthly analysis of the *energy waste* categories is demonstrated in Fig. 1.7. It represents, in percent, the number of submitted jobs, which results in categories I and II, related to the number of all active jobs. A considerable amount of jobs that are only processed by the scheduler and have a maximum running time of 30 s (category I)) represent from 14 to 56% of all submitted jobs throughout the whole period of investigation. On the contrary, jobs, which exceed the queue time limit, form less than 1% during the majority of the reported period. Seasonal steep rise of jobs' submission

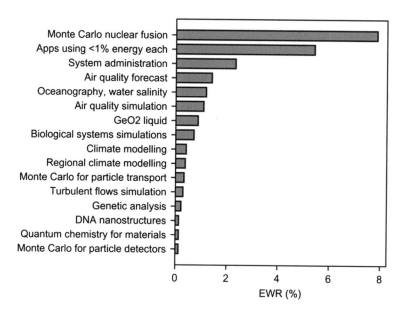

Fig. 1.6 Energy waste ratio (EWR) by applications

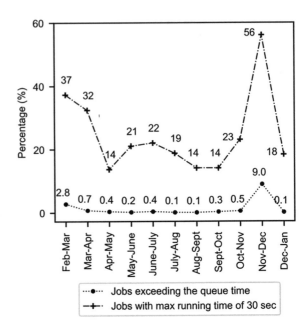

Fig. 1.7 Ratio of jobs causing *energy waste* because of processing time conflicts

from both categories happens in the month of November–December, which might be associated with users working from home during Christmas vacations and failing to conform to the DC policies.

To summarize, the results obtained through the assessment of *useful work* and *energy waste* reveal the energy consumption patterns within the cluster. First, the least energy is consumed during the summer months of annual vacations, whereas the most significant amount of wasted energy is observable in December–January when users might have worked remotely during the Christmas holidays. Second, a high percentage of jobs consume less than 10 kWh per month, which result in the energy spent on minor jobs rather than resource-hungry processes. Also, the cluster wastes most of the energy for jobs that end with errors for unknown reasons and that require further examination. Regarding the *energy waste* categories, some jobs that are only preprocessed by the scheduler and do not provide any results are considerably higher than the number of jobs removed from the queue because of the time limit conflicts.

1.4.6 Sustainability Analysis

DCs have progressively become subject to green audits, which entail an environmental impact assessment of all their IT and supporting facilities. It is partially caused by the fact that DC energy consumption indirectly implies the presence of carbon emissions associated with electricity generation and its utilization in DC for

Fig. 1.8 Monthly CO_2 (or equivalent) emissions caused by jobs, which ended with errors (dotted parts) and correctly finished jobs (inclined lines)

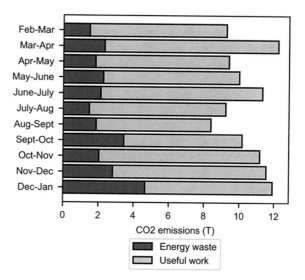

computational systems and DC supporting infrastructure. Therefore, inefficient use of energy has a negative impact on DC sustainability and thus on a smart city that supplies processing tasks to DC. Below we provide a way to assess this impact based on the available data of the cluster energy use. The idea is that the results in terms of energy consumption required for *useful work* and processes without positive results for the end user have been translated into CO_2 or equivalent greenhouse gases (GHG) emissions to show the environmental impact of the cluster's processing work.

In the context of the reported ENEA setting, a carbon factor is used to translate energy use of IT facilities operating under a specific workload to carbon and its equivalent emissions. Note that the carbon factor is taken from [31] for Italy. It converts electrical energy usage in MWh to tonnes of CO_2 or equivalent GHG and is equal to 0.343 tCO2/MWh.

Figure 1.8 shows the amount of carbon emissions produced by the computation facilities during the cluster processing. As a basis for this figure, the monthly energy consumption is calculated for all the jobs, which are successfully completed, and the jobs, which end up with errors. The values are converted to MWh and then multiplied by the carbon factor. The overall CO_2 emissions fluctuate between 8 and 12.2 tonnes CO_2 per month. The proportion of CO_2 emissions caused by *Not Useful Work* ranges from 16 to 40% of monthly emissions. Figure 1.8 is used here to highlight the importance of identifying jobs, which do not produce any *useful work*, but negatively impact the energy consumption and environment.

Evaluation of CO_2 or equivalent GHG emitted only by IT equipment does not facilitate the assessment of CUE metric for the cluster, because it requires data on total emissions caused both by IT equipment and supportive infrastructure of the DC. However, profiles of these emissions throughout the months of the monitoring campaign provide data to conduct a comparative analysis of DC's operations for

different periods and varying loads. This could be exploited to assess the effect of
the introduction of new policies. The current CUE metric estimation is impossible,
however, with the results depicted in Fig. 1.8. DC operators could assess a novel
metric, showing what portion of carbon emissions could have been produced for
jobs that have not been successfully completed and provide results to the end users.
Thus, by analogy with EWR, we propose to use Carbon Waste Ratio, CWR, to
express the same value in terms of CO_2 emissions:

$$CWR = \frac{\text{Carbon Emissions Introduced by Unproductive Computations}}{\text{Total DC Carbon Emissions over Time}}. \quad (1.8)$$

Here, we assume that the users of this metric possess a tool to estimate carbon
emissions of all DC system, which is not restricted to the example above, namely
the utilization of carbon factor to approximate carbon or equivalent emissions.

The conducted analysis provides a more in-depth insight into the *useful work* and,
at the same time, *waste energy* in the Cluster. An incremental contribution toward
a better understanding of DC sustainability has been presented in terms of carbon
emissions for useful work and jobs associated with *energy waste*. However, the data
available in the case study are not sufficient for the evaluation of any carbon or
sustainability metric.

Undeniably, sustainability metrics have formed a large set covering CUE or other
carbon emission and renewable energy use metrics. As proposed in this work, this set
could be completed by a metric evaluating indirect carbon emissions of the processes
associated with unproductive computations. It would emphasize the need for better
quality software products used by smart cities and facilitate a comparative analysis
of the DC performance in terms of environmental impact if there are changes in
software, workload, or any other component of the complex DC system.

1.5 Discussion

A quantitative assessment of the energy efficiency in DC Cluster and an overview
of computational processes discussed in the previous section provide fruitful ground
for the study of DC power management strategies [43]. Since the present work is
focused on computational efficiency and evaluation energy spent on the *useful work*,
the discussion raised in the current section is devoted to the elaboration of queuing
policies and possible application of best practices enabled by the previous results.

1.5.1   Energy Efficiency Benefits and Concerns of Jobs Execution in Parallel Mode

Parallel jobs are typically used to accelerate computations and increase the efficiency of resource usage. However, it is not infrequent that some jobs designed for serial execution are submitted to queues, which are designed to manage only parallel jobs. In addition, several other factors influence *energy waste* for *no work*, such as malfunctioning of the systems or algorithm errors.

In the specific studied scenario of CRESCO4 cluster operation, different LSF queues are configured for different kinds of jobs. In detail, serial queues are intended for executing jobs in serial mode only on one core, whereas parallel queues are designed for running jobs in a parallel mode so that jobs would run with multithreads on all the processor cores. When jobs not optimized for parallelism are submitted to the parallel queues, they cause big resource wastage. Among CRESCO4 applications, a common yet negative practice is observed when parallel jobs are poorly programmed, because they have a programming model better suited for serial jobs. As a result, when a job has been submitted to parallel queues with 24-processor core, for instance, and its specification states that the job could only run with the use of one processor core, consequently, the other 23 cores are left idle. This means more than 95% of resources have been wasted. Furthermore, the longer such jobs are being executed—the more energy is wasted. In addition to this, when such jobs are being run in parallel queues, some properly designed parallel jobs could have been withheld in the queue and remain in pending status and cumulatively cause *energy waste*.

A problem of resource wastage is also extended to the cases when users resubmit jobs with processing issues, either caused by the incorrect mode of execution stated before submission or other reasons. Namely, some jobs could have been dispatched and run, but no results have been generated due to problems with the executable, parameters or any options used. Usually, when users find processing issues associated with their jobs, they terminate and resubmit them. However, during this reiteration, both computing, energy resources of DC and users' time are wasted. A good strategy would be cooperating with users to test and debug programs and make sure they can run successfully and produce expected results before submitting them to the cluster's queues.

In essence, it would be greatly beneficial if appropriate codes are written and implemented for parallel jobs to exploit the real parallelization of computer resources. On the contrary, jobs are not well parallelized generally waste resources. Response time and results acquisition would take longer, and such jobs would have adverse effects on other processes in such cases. DC operators could introduce additional evaluation of a proper parallelization of processes. A generally accepted formula is CPU Time Consumed divided by Turnaround Time minus Waiting Time in the queue (e.g., [CPU Time/(Turnaround Time—Waiting Time)]). When the value obtained by the equation approaches the number of cores used, then the parallel algorithm can be considered appropriate.

1.5.2 Data Center Energy Efficiency Policies and Strategies

This subsection aims to generalize DC energy efficiency policies and strategies starting to the results obtained in our specific case study. The DC Cluster case study described previously implies that some improvements in power management could be useful for the DC in question. Thus, we summarize the results obtained through analysis of monitoring data gathered in DC to make suggestions on the power management strategies. The proposals aim to reduce energy consumption and subsequently, to improve the energy efficiency of DC. For this purpose, three fundamental aspects of DC resource manipulation are distinguished: (1) scheduling, (2) resource allocation and (3) resource management.

Scheduling is responsible for defining the order of execution for the active tasks, and as aforementioned, the strategy used in the target DC cluster is based on the First Come First Served (FCFS) algorithm. However, considering the obtained results and the capabilities of the LSF platform in terms of scheduling, the order in which tasks are placed in the queues could be defined with other strategies. Some widely used algorithms and related policies explaining the scheduling of tasks in DCs have been discussed in [21]. For example, the Last Come First Served (LCFS) could be used instead of FCFS; furthermore, the Largest Job First (LJF) aims to optimize the utilization of the system, the Smallest Job First (SJF) increases the throughput of the system. Scheduling algorithms could be also supplemented with a backfilling approach. When the backfilling is applied, the positions of the jobs in the queue are changed based on the availability of the resources and the priorities of the tasks. Some examples of backfilling techniques are Aggressive Backfill, Relaxed Backfill, Conservative Backfill.

The resource allocation strategy provided to the users of the cluster is based on the combination of "free queue utilization" and the "on time" approaches: the first user who enters the system uses the first available queue with requested characteristics. However, since the tasks are submitted by different users over time, the decision of where to execute each arriving task is usually made in an online manner without knowledge of the future task arrivals. The performance-energy trade-off requires the allocation strategy to minimize both the average task response time and the overall energy consumption. Another approach could be to assign tasks in a way that consolidates the workload in a predefined manner. For example, instead of using a simple strategy of allocating the jobs to the first available resource (node), resource allocation could be improved by analyzing first the temperature distribution and energy use so that the jobs will be allocated to the nodes optimal in terms of energy consumption and performance.

Resource management policies could be categorized into three groups: management of virtual resources, physical machines and applications. The first two groups usually require support from the underlying hardware layer and their effectiveness is closely related to the IT equipment. The first category comprises strategies for VM allocation, a topic that is currently the focus of researchers and industries but lies beyond the scope of this work. The most popular policies for the second category, which focuses on physical machine usage and energy efficiency, are the Switching nodes ON/OFF and the Dynamic Voltage and Frequency Scaling (DVFS). However, due to the unknown pattern of the future load, it might be challenging to identify which nodes are optimal for switching into the OFF mode and the appropriate duration of the inactive mode. The current work does not aim at predicting the loads, but as a side benefit of analyzing energy consumption of the applications and, specifically, their energy use profiles throughout the months could be used in prediction algorithms and optimize the switch ON/OFF policy in practice. The third category of resource utilization by applications is explicitly studied in the quantitative analysis of energy consumption by applications.

According to the CRESCO4 analysis results, scheduling, resource allocation and resource management strategies can be combined in ad-hoc policies for power management. There are several ways to follow to obtain "workload-driven" power management. First of all, the operator of the target DC prioritizes power cost optimization. From LSF information, it can be seen that the majority of job requests arrives in the morning during working hours. Generally, it is the time when energy cost is high due to power cost changes throughout the day. High demand causes a high price in the energy market and vice-versa, low demand suggests a low price. The power cost optimization strategy could be used to minimize power cost by shifting the workload to low-cost periods, when it is possible or when the company policies allow it. In the case where system utilization is close to 100%, an alternative strategy is to schedule only high priority work during the highest power cost period and prevent workload which can wait (i.e. low priority) from consuming power when it is most expensive.

From the results obtained, it is observed that the system is not always 100% loaded. In contrast, the system is generally loaded only by half or even less. It is also deduced that the goal of concentrating the load is not pursued because waiting time for the jobs is minimized in the first place leaving the energy consumption minimization as a lower priority. The power consumption of a node depends on its operational mode. When the node is off, and in sleep/standby mode, it consumes the least power. If a node is idle, it consumes 40–50% of power compared to a fully loaded node. For the power efficiency optimization strategy to work, the understanding of a particular class of server's performance per kilowatt is initially required. Such a metric might be application-dependent and, therefore, should be considered carefully. Using tools such as Platform LSF, one can easily manage the distribution of a load on the various nodes. During the analysis, some applications are detected to have required intensive I/O or, other applications that wait for MPI messages. These applications can be classified as low load or cooling load because the CPU tends to go idle and consume less power during these waiting periods. The strategies mentioned above could be replicated in other DCs with similar configurations; hence, it is possible to redistribute the jobs to optimize the energy consumption and also consider the thermal operational characteristics [13].

1.5.3 Sustainability-Oriented DC

The DC cluster experiences peaks of energy consumption in the morning hours, which, apart from increasing energy cost, affect the amount of carbon emitted as a consequence of electricity generation and further utilization by the DC. Two ways of reducing the harmful impact of energy consumption peaks might be brought into action: (1) executing only "high priority work during the highest power cost period" and (2) equip DC with it owns sources of renewable energy such as solar panels and windmills to further adapt the execution of jobs according to the availability of local renewable energy production [9].

Further consideration of the sustainable issue shows that the ISO standard 30,134 being developed for sustainable IT and DCs includes a renewable energy factor as a key performance indicator for DCs [32]. However, other ISO documents regarding IT, energy efficiency and sustainable topics in DC are currently under development. Even if the ongoing innovation actions (Reduce/Reuse/Recycling) help to support a DC sustainability, poor utilization of the defined metrics related to IT and *useful work* will impede business innovation and pose as a barrier to meeting environmental sustainability goals. Also, for this reason, in this work, the authors have investigated the real practical application of the EWR metric. In summary, the analysis precisely pinpoints the non-computing and *Not Useful Work* within *energy waste*. Therefore, based on the knowledge of *energy waste*, it is possible to improve the power management and pursue DC sustainability in relation to IT jobs energy efficiency.

1.6 Conclusion

This chapter focuses on DC energy efficiency required by a smart city that aims to achieve sustainability goals. DCs contribute to the smart city as enablers of its urban technologies, because they process big data coming from the city's interconnected systems and as a result of societal use of applications. In addition, with the introduction of the IoT in the urban environment, it is witnessing an explosive proliferation of endpoints, and consequently, of a continuous flow of the heterogeneous data produced by various applications. Under these circumstances, DCs are required to provide computational, storage and network resources in a reliable way to ensure safety and comfort of the citizens through continuous availability of shared resources. At the same time, DCs must operate efficiently to minimize capital and operational expenditure and indirect environmental impact of emissions associated with energy production and use within the DC. The challenges of DC operation for smart city also comprise efficient cooling, managing advanced virtualization techniques, power and IT equipment energy efficiency, development of a holistic framework based on a common set of metrics for energy efficiency and sustainability assessment, standardization of metrics, use of common guidelines. From this list of needed improvements, the chapter has addressed an issue of measuring and achieving high computational productivity and low energy waste for IT operation.

The added value of this work is that DC energy efficiency is addressed from a smart city point of view. The authors have proposed a methodology to address the issue of energy productivity through the consideration of computational processes energy efficiency within a DC cluster and have applied it to a real dataset of ENEA DC Cluster power use and scheduler reports. Analysis of computational energy productivity is conducted with the metrics EWR and DCeP to evaluate energy spent on *useful work* and *energy waste* on incorrectly finished computational processes. Further insight into a DC environmental impact is provided by an assessment of the DC indirect carbon emissions associated with *useful work* and *energy waste*. This assessment has resulted in a new metric proposal, Carbon Waste Ratio, that measures carbon emissions introduced by the processing of jobs that have not provided any results for the users who have submitted their jobs.

This work is aimed at various DC and smart city operators who are motivated to apply energy efficiency strategies in their DCs and minimize environmental impact in the smart cities. Thus, the example of energy productivity metrics practical use for a real DC discussed in this chapter could be a recommendation for DC operators to assess their DC computational productivity and related carbon emissions. This assessment might further be used by a smart city to evaluate direct/indirect carbon emissions as well as savings through DC processing of integrated urban applications. Moreover, Intelligent Data Analytics could provide valuable insight into patterns of user or entities (e.g., smart vehicles, smart buildings) engagement with smart city applications (e.g., smart energy, smart traffic, smart grid, smart health). Through this, a predictive engagement model could be built in order to render a stream of DC activities more predictable. Such a model could facilitate a more proactive form of task management within a DC and facilitate better management of task scheduling and better adaptation of DC activities to locally sourced green energy for reducing carbon emission. In addition, the proposed methodology could be reused with adjustments in the developing area of edge computing where computational productivity is equally important, although the system architecture differs from the one that only uses DC for processing the smart city data. Inevitably, the activities on energy efficiency assessment and possible improvements of smart cities will affect the quality of life of its residents. This will enable them to gain access to applications, which operate optimally and waste less energy.

Acknowledgements The research work has been supported and funded by the PERCCOM Erasmus Mundus Program of the European Union [35]. Moreover, the authors would like to express their gratitude to the research HPC group at the ENEA-R.C. Portici for the useful advice on modeling and control of ENEA-Data Center.

References

1. Acton M, Bertoldi P, Booth J, Newcombe L, Rouyer A, Tozer R (2018) 2018 best practice guidelines for the eu code of conduct on data centre energy efficiency, v.9.1.0. Available via DIALOG. http://publications.jrc.ec.europa.eu/repository/bitstream/JRC110666/kjna29103 enn.pdf. Accessed 27 March 2019
2. Al Nuaimi E, Al Neyadi H, Mohamed N, Al-Jaroodi J (2015) Applications of big data to smart cities. J Internet Serv Appl 6: 1–15. https://doi.org/10.1186/s13174-015-0041-5
3. Allam Z, Dhunny Z (2019) On big data, artificial intelligence and smart cities. Cities 89: 80–91. https://doi.org/10.1016/J.CITIES.2019.01.032
4. Alshawish RA, Alfagih SAM, Musbah MS (2016) Big data applications in smart cities. In: 2016 international conference on engineering & MIS (ICEMIS), pp 22–24. https://doi.org/10.1109/ICEMIS.2016.7745338
5. Antal M et al (2018) Transforming data centers in active thermal energy players in nearby neighborhoods. Sustainability 10(4):939. https://doi.org/10.3390/su10040939
6. ASHRAE Technical Committee 9.9, 2011 (2011) Thermal guidelines for data processing environments—expanded data center classes and usage guidance. American Society of Heating, Refrigerating, and Air-Conditioning Engineers Inc.
7. ASHRAE (2016) Data center power equipment thermal guidelines and best practices, Technical Commitee 9.9 of American Society of Heating, Refrigeration and Air Conditioning Engineering
8. Azevedo D et al (2010) The green grid: carbon usage effectiveness (CUE): a green grid data center sustainability metric. White Paper #32, The Green Grid, Beaverton, O, USA. https://airatwork.com/wp-content/uploads/The-Green-Grid-White-Paper-32-CUE-Usage-Guidelines.pdf. Accessed 29 March 2019
9. Beldiceanu N et al (2017) Towards energy-proportional clouds partially powered by renewable energy. Computing 99(1):3–22. https://doi.org/10.1007/s00607-016-0503-z
10. Bibri SE (2018) 'The IoT for smart sustainable cities of the future: an analytical framework for sensor-based big data applications for environmental sustainability. Sustain Cities Soc. Elsevier 38: 230–253.https://doi.org/10.1016/J.SCS.2017.12.034
11. Capozzoli A et al (2014) Thermal metrics for data centers: a critical review. Energy Procedia. Elsevier 62: 391–400.https://doi.org/10.1016/J.EGYPRO.2014.12.401
12. Capozzoli A et al (2015) Review on performance metrics for energy efficiency in data center: the role of thermal management. Lecture notes in computer science (including subseries lecture notes in artificial intelligence and lecture notes in bioinformatics) 8945:135–151. https://doi.org/10.1007/978-3-319-15786-3_9
13. Capozzoli A, Primiceri G (2015) Cooling systems in data centers: state of art and emerging technologies. Energy Procedia 83:484–493. https://doi.org/10.1016/j.egypro.2015.12.168
14. Cioara T et al (2019) Exploiting data centres energy flexibility in smart cities: business scenarios. Inf Sci 476:392–412. https://doi.org/10.1016/j.ins.2018.07.010
15. Chen H, Chiang RHL, Storey VC (2012) Business intelligence and analytics: from big data to big impact. MIS Q: Manage Inform Syst 36(4):1165–1188
16. Chinnici M, Quintiliani A (2013) An example of methodology to assess energy efficiency improvements in datacenters. In: 2013 international conference on cloud and green computing. IEEE, Karlsruhe, Germany, pp 459–463. https://doi.org/10.1109/CGC.2013.78
17. Chinnici M, Capozzoli A, Serale G (2016) Measuring energy efficiency in data centers. In: Pervasive computing: next generation platforms for intelligent data collection, pp 299–351. https://doi.org/10.1016/B978-0-12-803663-1.00010-3
18. Chinnici M, De Chiara D, Quintiliani A (2017) Data center, a cyber-physical system: improving energy efficiency through the power management. In: 2017 IEEE 15th international conference on dependable, autonomic and secure computing, 15th international conference on pervasive intelligence and computing, 3rd international conference on big data intelligence and computing and cyber science and technology congress (DASC/PiCom/DataCom/CyberSciTech). IEEE, Orlando, FL, USA, pp 269–272. https://doi.org/10.1109/DASC-PICom-DataCom-CyberS ciTec.2017.56

19. Chinnici M, De Chiara D, Quintiliani A (2017) An HPC-data center case study on the power consumption of workload. In: Ntalianis K, Croitoru A (eds) Lecture notes in electrical engineering book series (LNEE, vol 489). Springer, Cham, pp 183–192. https://doi.org/10.1007/978-3-319-75605-9_26
20. Chinnici M, De Vito S (2018) IoT meets opportunities and challenges: edge computing in deep urban environment. In: Kharchenko V, Kor AL, Rucinski A (eds) Dependable IoT for human and industry. Modeling, architecting, implementation. River Publishers Series in Information Science and Technology
21. Cupertino L et al (2015) Energy-efficient, thermal-aware modeling and simulation of data centers: the CoolEmAll approach and evaluation results. Ad Hoc Netw 25:535–553. https://doi.org/10.1016/j.adhoc.2014.11.002
22. Data Center Equipment (2019) In: Energystar.gov. https://www.energystar.gov/products/data_center_equipment. Accessed 21 Mar 2019
23. Davies GF, Maidment GG, Tozer RM (2016) Using data centres for combined heating and cooling: an investigation for London. Appl Therm Eng 94:296–304. https://doi.org/10.1016/j.applthermaleng.2015.09.111
24. Dc4cities.eu (2019) DC4Cities|Adapt and be Adapted. http://www.dc4cities.eu/en/adapt-being-adapted/. Accessed 28 Mar 2019
25. Ebrahimi K, Jones GF, Fleischer AS (2014) A review of data center cooling technology, operating conditions and the corresponding low-grade waste heat recovery opportunities. Renew Sustain Energy Rev. https://doi.org/10.1016/j.rser.2013.12.007
26. Gandomi A, Haider M (2015) Beyond the hype: big data concepts, methods, and analytics. Int J Inf Manage 35:137–144. https://doi.org/10.1016/j.ijinfomgt.2014.10.007
27. Garraghan P, Moreno IS, Townend P, Xu J (2014) An analysis of failure-related energy waste in a large-scale cloud environment. IEEE Trans Emerg Topics Comput 2: 166–180. https://doi.org/10.1109/TETC.2014.2304500
28. Grishina A et al (2018) DC energy data measurement and analysis for productivity and waste energy assessment. In: 2018 IEEE international conference on computational science and engineering (CSE). IEEE, Bucharest, Romania, pp 1–11. https://doi.org/10.1109/CSE.2018.00008
29. Grishina A, Chinnici M, De Chiara D, Rondeau E, Kor A (2018) Energy-oriented analysis of HPC cluster queues: emerging metrics for sustainable data center. Lecture Notes in Electrical Engineering. ISSN 1876–1100 (in press)
30. Hashem IAT et al (2016) The role of big data in smart city. Int J Inf Manage 36(4):1165–1188. https://doi.org/10.1016/j.ijinfomgt.2016.05.002
31. Intel (2010) Increasing data center efficiency with server power measurements, White Paper. https://www.intel.com/content/dam/doc/white-paper/intel-it-data-centerefficiency-%0Aserver-power-paper.pdf. Accessed 30 July 2018
32. ISO (2019) Data centres. https://www.iso.org/search.html?q=data%20centres. Accessed 27 Mar 2019
33. Khan MAUD, Uddin MF, Gupta N (2014) Seven V's of big data understanding big data to extract value. In: Proceedings of the 2014 zone 1 conference of the american society for engineering education— "Engineering Education: Industry Involvement and Interdisciplinary Trends", ASEE Zone 1 2014. IEEE Computer Society. https://doi.org/10.1109/ASEEZone1.2014.6820689
34. Khosravi A, Garg SK, Buyya R (2013) Energy and carbon-efficient placement of virtual machines in distributed cloud data centers. In: Lecture notes in computer science (including subseries lecture notes in artificial intelligence and lecture notes in bioinformatics), pp 317–328. https://doi.org/10.1007/978-3-642-40047-6_33
35. Klimova A et al (2016) An international Master's program in green ICT as a contribution to sustainable development. J Clean Prod 135:223–239. https://doi.org/10.1016/j.jclepro.2016.06.032
36. Klingert S, Chinnici M, Porto MR (2014) Preface of lecture notes in computer science (including subseries lecture notes in artificial intelligence and lecture notes in bioinformatics), vol 8945, pp V–VII. ISSN: 03029743, ISBN: 978–331915785–6

37. Lim C, Kim K-J, Maglio PP (2018) Smart cities with big data: Reference models, challenges, and considerations. Cities 82:86–99. https://doi.org/10.1016/j.cities.2018.04.011
38. Munteanu I et al (2013) Efficiency metrics for qualification of datacenters in terms of useful workload. In: 2013 IEEE grenoble conference. IEEE, Grenoble, France, pp 1–6. https://doi.org/10.1109/PTC.2013.6652470
39. Neirotti P et al(2014) Current trends in smart city initiatives: some stylised facts. Cities. Pergamon, 38: 25–36.https://doi.org/10.1016/J.CITIES.2013.12.010
40. Osman AMS (2019) A novel big data analytics framework for smart cities. Futur Gener Comput Syst 91:620–633. https://doi.org/10.1016/j.future.2018.06.046
41. Pattinson C et al (2014) Green sustainable data centres, measurement and control. Available via DIALOG. https://www.ou.nl/documents/380238/382808/GSDC_05_Measurement_and_control.pdf. Accessed 29 Mar 2019
42. Patterson M, Azevedo D, Belady C, Pouchet J (2011) Water Usage Effectiveness (WUE)—a green grid data center sustainability metric. White Paper #35, The Green Grid, Beaverton, O, USA. https://airatwork.com/wp-content/uploads/The-Green-Grid-White-Paper-35-WUE-Usage-Guidelines.pdf. Accessed 29 Mar 2019
43. Postema BF, Haverkort BR (2018) Evaluation of advanced data centre power management strategies. Electron Notes Theor Comput Sci 337:173–191. https://doi.org/10.1016/j.entcs.2018.03.040
44. Quintiliani A, Chinnici M, De Chiara D (2016) Understanding "workload-related" metrics for energy efficiency in data center. In: 2016 20th international conference on system theory, control and computing (ICSTCC). IEEE, Sinaia, Romania, pp 830–837. https://doi.org/10.1109/ICSTCC.2016.7790771
45. Quintiliani A, Chinnici M (2016) D7.3—final DC4Cities standardization framework and results description of the European Cluster. Rome, Italy. http://www.dc4cities.eu/en/wp-content/uploads/2016/05/D7.3-Final-DC4Cities-standardization-framework-and-results-description-of-the-European-Cluster.pdf
46. Royaee, Z., Mohammadi M. (2013) Energy aware Virtual Machine Allocation Algorithm in Cloud network. SGC2013 Smart Grid Conference 17–18 Dec. 2013., pp. 259–263. doi: https://doi.org/10.1109/SGC.2013.6733819.
47. Reddy VD et al (2017) Metrics for sustainable data centers. IEEE Trans Sustain Comput 2(3):290–303. https://doi.org/10.1109/TSUSC.2017.2701883
48. Barns S (2016) Mine your data: open data, digital strategies and entrepreneurial governance by code. Urban Geogr 37(4):554–571. https://doi.org/10.1080/02723638.2016.1139876
49. Wahlroos M, Pärssinen M, Manner J, Syri S (2017) Utilizing data center waste heat in district heating—impacts on energy efficiency and prospects for low-temperature district heating networks. Energy 140: 1228–1238. https://doi.org/10.1016/j.energy.2017.08.078
50. Zakarya M (2018) Energy, performance and cost efficient datacenters: a survey. Renew Sustain Energy Rev 94:363–385. https://doi.org/10.1016/j.rser.2018.06.005

Chapter 2
Apache Spark for Digitalization, Analysis and Optimization of Discrete Manufacturing Processes

Dorin Moldovan, Ionut Anghel, Tudor Cioara, and Ioan Salomie

Abstract Digitalization, analysis and optimization of discrete manufacturing processes represent a research challenge because the data generated by the sensors that monitor the manufacturing processes are characterized by noise, missing values, many features due to the fact that the assembly processes involve many different assembly steps and often the resources that are used in the manufacturing processes are applied inefficiently. Digitalization of assembly processes using the latest technologies, analysis of data generated by the monitoring sensors using big data technologies and optimization of the manufacturing processes by identifying the steps that have the highest impact on the final output represent serious research challenges, and in this chapter, we approach these challenges by: (1) creating a platform in which the users can visualize the assembly steps of the discrete manufacturing processes, (2) identifying the products that have manufacturing faults using machine learning and deep learning models and (3) optimizing the parameters of the machine learning and deep learning models using bio-inspired heuristics.

Keywords Discrete manufacturing processes · Machine learning · Deep learning · Sensors · Internet of things

D. Moldovan · I. Anghel (✉) · T. Cioara · I. Salomie
Computer Science Department, Technical University of Cluj-Napoca, Cluj-Napoca, Cluj, Romania
e-mail: ionut.anghel@cs.utcluj.ro

D. Moldovan
e-mail: dorin.moldovan@cs.utcluj.ro

T. Cioara
e-mail: tudor.cioara@cs.utcluj.ro

I. Salomie
e-mail: ioan.salomie@cs.utcluj.ro

F. Pop and G. Neagu (eds.), *Big Data Platforms and Applications*,
Computer Communications and Networks,
https://doi.org/10.1007/978-3-030-38836-2_2

2.1 Introduction

Manufacturing processes are characterized by a high complexity due to the fact that the manufacturing of the products requires a large sequence of manufacturing steps that must be followed to obtain the final products, it is necessary to have a perfect coordination between the resources that are responsible for the manufacturing of the products and all the pieces that compose the final products must be assembled correctly because otherwise the final products are faulty and have no practical use. The datasets that characterize the manufacturing processes are usually characterized by a big number of features due to the big number of steps through which the products must pass from start to end and also due to the big number of sensors that might monitor the manufacturing processes. The data are also characterized by noise and missing values that might be caused by the environment through which the data must pass from the moment it is recorded until the moment it is used for analysis.

In this article, the main focus is the analysis of discrete manufacturing processes and more specifically the analysis of the products that are the outcome of the manufacturing processes using as illustrative use case the manufacturing of the regulators in Emerson factory. The regulators are composed of several components such as regulator shutters, travel indicators and inlet and outlet flanges. Each component is further composed of several materials. The regulators follow a series of assembly steps that must be executed without errors in order to obtain high-quality products with no faults. The complexity of the manufacturing of the FL series regulators is, thus, given by the big number of assembly steps that must be performed. Using novel approaches from machine learning and deep learning areas, we aim to determine the causes that might lead to products of low quality and to optimize the manufacturing processes.

The research results presented in this article are in the context of the project OptiPlan [1], which proposes novel technologies for digitalization, analysis and optimization of production processes of gas regulators in Emerson factory.

The main contributions of the article are:

- the development of a platform for digitalization of discrete manufacturing processes;
- the analysis of the faults associated with the discrete manufacturing processes using a machine learning methodology developed in-house;
- the optimization of the discrete manufacturing processes using bio-inspired heuristics;
- the application of the proposed technologies on two datasets from literature and on one synthetic dataset using an experimental prototype that was developed in-house.

The rest of the article is organized as follows: Sect. 2.2 presents the background, Sect. 2.3 presents the main materials and methods, Sect. 2.4 presents the results, Sect. 2.5 presents a discussion of the results and Sect. 2.6 presents the conclusions.

2.2 Background

This section is organized into five major parts: (1) the first part presents an overview of several approaches that are used in literature for smart manufacturing processes which are based on IoT techniques, (2) the second part presents a synthesis of the machine learning approaches that are used in the literature for the analysis of manufacturing processes, (3) the third part presents literature approaches that are used for the optimization of manufacturing processes, (4) the fourth part presents a synthesis of the bio-inspired techniques that are used in the literature for tuning the parameters of the machine learning models and (5) the fifth part presents approaches used in our research for the analysis of the faults in manufacturing processes.

2.2.1 IoT for Smart Manufacturing Processes

As technologies that are used in manufacturing processes advance, the manufacturing processes become more sophisticated. Industry 4.0 concept is introduced in [2] and one of its major objectives is the creation of smart factories through which goods of higher quality can be produced at a lower cost. In [3], it is presented a smart manufacturing platform that is based on the following technologies: big data analytics, IoT, cloud computing and cyberphysical systems. The IoT can be adopted as a data collection and communication platform for smart factories in Industry 4.0 era. Cloud computing can be used to create a variety of cloud manufacturing services that can be used to support different manufacturing activities such as the design of the products, the simulation of the manufacturing processes, the manufacturing of the products, the testing of the products, the management of the resources and all other tasks that might exist in the life cycle of the products [4]. The authors of [5] provide a concept for intelligent network services in the case of flexible smart manufacturing, which considers the identified needs, while the authors of [6] propose a demand response scheme that is based on the Stackelberg model for smart manufacturing systems that have a massive size. In [7], the authors develop a manufacturing system that is IoT enabled for tool wear characterization and their main contribution is the design of an IoT architecture for a tabletop Computer Numeric Control (CNC) milling machine.

2.2.2 Machine Learning Approaches for Manufacturing Process Analysis

The authors of [8, 9] apply several classification algorithms on the semiconductor manufacturing process dataset (SECOM) [10] such as: Logistic Regression (LR), Decision Trees (DT) and Artificial Neural Network (ANN) in order to detect faults in the products, which are the results of the manufacturing processes. These classification algorithms are used together with oversampling techniques such as SMOTE (Synthetic Minority Over-sampling Technique), which improve their performance. In [11], the semiconductor tool fault isolation (SETFI) dataset [12] is used as experimental support for the following classification algorithms: Decision Trees (DT), Random Forest (RF), Naive Bayes (NB), k-Nearest Neighbor (k-NN), Artificial Neural Network (ANN) and Support Vector Machine (SVM). The presented sequence analysis method identifies tool patterns from the processing data that are characteristic for wafer fabrication. The patterns that are extracted using the proposed approaches can differentiate the processes that have a high performance from the low-performance processes. In [13], several deep learning algorithms such as Artificial Neural Network (ANN) and Fuzzy Neural Network (FNN) are applied on a dataset that was generated at 6 inch semiconductor wafer fabrication system (SWFS) in Shanghai in order to approach the optimization of the rescheduling strategy. The authors of [14] apply the Support Vector Machine (SVM) algorithm on data from an abrasion-resistant material manufacturing process for predicting the quality of the manufacturing process; while in [15], the data from an electronic chip manufacturing process are analyzed using an Artificial Neural Network (ANN). In [16], the authors propose a fault detection algorithm that combines three phases: the extraction of the features, the selection of the features and the classification. The algorithm is applied to a semiconductor process dataset. The two feature selection algorithms that are applied in the features selection phase are the Principal Component Analysis (PCA) and the Support Vector Machine Recursive Feature Elimination (SVM-RFE). The approach that is presented in [17] uses the Randomized K-d Tree ReliefF algorithm for the selection of the features and is tested on two datasets from industrial processes. The features are selected in [18] using a genetic algorithm (GA). The fitness function of the genetic algorithm is developed based on a Hidden Markov Model (HMM) and the approach is evaluated on data from semiconductor wafer production equipment that was recorded during a period of 1 year. The developed approach repeats the choosing of the feature subsets, the training of the HMM and the evaluation of the performance until convergence. In [19], the authors propose a diagnosis system architecture in order to evaluate the faults from imbalanced and incomplete data in the case of smart semiconductor manufacturing processes, and in [20], it is developed a new data analytics framework that emphasizes the importance of the preprocessing phase in learning artificial intelligence (AI) models for complex manufacturing systems.

2.2.3 Manufacturing Processes Optimization Literature Approaches

In [21], the authors propose a manufacturing process optimization system which is based on Quantum Particle Swarm Optimization (QPSO), and in [22], the authors describe a framework that is used for the optimization of electronic manufacturing processes, which consists of three phases: (1) a system-modeling phase, (2) a system analysis and control phase and (3) a system optimization phase. In [23], it is proposed a simulation-based system for the optimization of manufacturing processes, which is addressed by the issue of process variability; the authors of [24] address challenges such as electrical performance, reliability and peripheral interfacing in the case of printed electronics manufacturing process optimization, and in [25], it is described an approach for the optimization of the manufacturing processes, which is based on rough theory. The authors of [26] propose a data mining approach for the optimization of the manufacturing processes in which they consider three important factors, namely, flexible manufacturing, information about final customer feedback and supplier selection. In [27], it is proposed a method for semiconductor process optimization that is based on the use of functional representations of spatial variation, and in [28], the authors propose a method for semiconductor manufacturing processes optimization that consists of the following steps: (1) electroplating process step, (2) plasma ashing step and (3) metal insulator metal capacitor step.

2.2.4 Bio-Inspired Techniques for Tuning the Parameters of Machine Learning Models

In this subsection, there are presented a selection of bio-inspired techniques that are used in literature for tuning the parameters of machine learning models. The authors of [29] apply the Particle Swarm Optimization (PSO) algorithm in order to optimize the parameters of deep learning models, and in [30], the Ant Colony Optimization (ACO) algorithm is used for the optimization of deep recurrent neural networks. Chicken Swarm Optimization (CSO) is used in [31] for features selection, Bat Algorithm (BA) is used in [32] for tuning extreme machine learning models, Centripetal Accelerated Particle Swarm Optimization (CAPSO) is applied in [33] for artificial neural network learning enhancement, Artificial Fish Swarm Algorithm (AFSA) is applied in [34] for improving and learning of an artificial neural network (ANN) model and Harmonious Cat Swarm Optimization (HCSO) is applied in [35] for feature selection in the case of support vector machine (SVM).

2.2.5 Approaches Used in Our Research for the Analysis of the Faults in Manufacturing Processes

The analysis of the faults in manufacturing processes was approached by us in [36–38] where we applied machine learningand deep learning techniques on the datasets SECOM and SETFI that are illustrative for manufacturing processes in order to detect if the resulted products are faulty or not faulty, but compared with the methods presented in those articles, in this article: (1) we introduce a novel synthetic dataset that is illustrative for the manufacturing of the FL series regulators, (2) we tune the parameters of the machine learning models using approaches that are more complex and (3) we propose a future research direction that is based on semantic web technologies.

2.3 Materials and Methods

This section presents a synthesis of the main materials and methods used for digitalization, analysis and optimization of the discrete manufacturing processes.

2.3.1 Architectural Prototype for Simulating the Manufacturing of FL Series Regulators

Figure 2.1 presents the high-level view of the architectural prototype that is used for simulating the data that is generated by the manufacturing of the FL series regulators.

The architectural prototype has seven major components: (1) a front-end application created in angular that is used by the end-users for the interaction with the back-end and which is deployed on NGINX, (2) a back-end application created in Spring Boot that is deployed on Tomcat, (3) a component that uses the Zookeeper and the

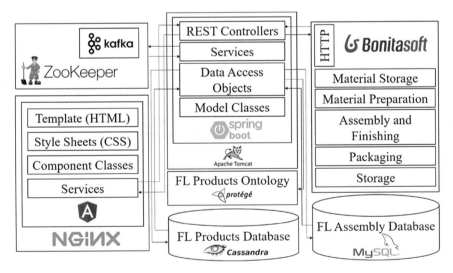

Fig. 2.1 High-level view of the architectural prototype used for generating synthetic data for the FL series regulators

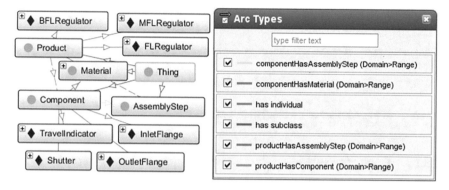

Fig. 2.2 FL series regulators ontology

Kafka servers for simulating the real-time transmission of the data from the sensors that monitor the manufacturing of the products, (4) an application created using the Bonita software for simulating the steps that are followed in the creation of the products, (5) a MySQL database that is used for the persistence of the parameters and of the configurations used for the assembly of the products, (6) an ontology created in Protege that describes the products used in the simulation and (7) a Cassandra database that persists the data about the final products.

The ontology that is used as support for generating the FL regulator synthetic dataset used in our simulation is presented in Fig. 2.2.

In our simulation we suppose that each product is composed of several components and each component is composed of several materials. We define a product as:

$$product = [component_1, \ldots, component_N] \qquad (2.1)$$

where N is the number of components and the generic structure of a component is:

$$component = [material_1, \ldots, material_M] \qquad (2.2)$$

where M is the number of materials. Next are presented the procedures, which are used for the assembly of: (a) materials, (b) components and (c) products.

2.3.1.1   Materials Assembly

The assembly time for the ith material is defined as:

$$a_t(material_i) = Random[\min(a_t(material_i)), \max(a_t(material_i))] \qquad (2.3)$$

where $\min(a_t(material_i))$ is the minimum assembly time for the ith material and $\max(a_t(material_i))$ is the maximum assembly time for the ith material.

2.3.1.2 Components Assembly

Each component follows a series of assembly substeps as presented next:

$$componentassembly = [substep_1, \ldots, substep_K] \qquad (2.4)$$

where K is the number of assembly substeps.

2.3.1.3 Products Assembly

Each product follows a series of assembly steps that are presented next:

$$productassembly = [step_1, \ldots, step_L] \qquad (2.5)$$

where L is the number of assembly steps.

Using the presented architectural prototype in the context of the project OptiPlan [1], we created one synthetic dataset: the FL regulators synthetic dataset. Next are presented examples of assembly steps that must be followed in the assembly of the FL series regulators: (1) lubricate the channel, (2) insert the rubber seal, (3) lubricate the inside channel of the flange, (4) insert the antifriction rings separated by the rubber seal and (5) screw the travel indicator subassembly in the cover.

2.3.2 Machine Learning Methodology for Detecting Faulty Products in Discrete Manufacturing Processes

In Fig. 2.3, it is presented the methodology that is used for detecting faulty products in discrete manufacturing processes.

The phases that are used in the machine learning methodology proposed by us are described in more detail next.

2.3.2.1 Discrete Manufacturing Processes Datasets

The three manufacturing process datasets that are used in order to test and to validate the proposed machine learning methodology are: (1) the FL regulators synthetic dataset [1], (2) the semiconductor manufacturing process dataset (SECOM) [10] and (3) the semiconductor tool fault isolation (SETFI) dataset [12]. A part of these datasets that are used in experiments are processed prior to the application of the machine learning methodology in order to simulate discrete manufacturing processes.

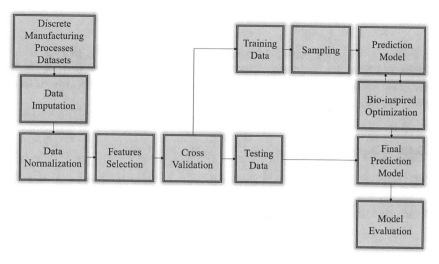

Fig. 2.3 Machine learning methodology for detecting faulty products in discrete manufacturing processes

2.3.2.2 Data Imputation Phase

This is the phase in which the missing values are replaced. Even though in the literature there are various heuristics that treat the replacement of the missing values such as: the median heuristic, the most frequent value heuristic, the next value heuristic and the linear interpolation heuristic, in our approach, we chose to apply the mean heuristic because this is one of the simplest heuristics and because the data generated by the sensors should be a continuous function.

2.3.2.3 Data Normalization Phase

In this phase, the data are normalized in order to take values from the interval [0, 1]. This operation is necessary in order to prepare the data prior to the application of the machine learning prediction models because the machine learning models require the input data to be in the interval [0, 1].

2.3.2.4 Feature Selection Phase

The Principal Component Analysis (PCA) algorithm that is used for the selection of the features of the discrete manufacturing processes datasets is based on the classical PCA algorithm [39].

2.3.2.5   Cross-Validation Phase

In the cross-validation phase, the data are split into two datasets, namely, a train dataset (80%) and a test dataset (20%). The train dataset is used for training the prediction models and the test dataset is used for evaluating the performance of the prediction models.

2.3.2.6   Sampling Phase

This phase oversamples the training data in order to generate more samples, which can be used as input for the prediction model. The data are oversampled using SMOTE (Synthetic Minority Over-sampling Technique) [40].

2.3.2.7   Prediction Model Construction Phase

The two prediction models that are applied in this chapter are: (1) a Random Forest (RF)-based prediction model and (2) a Neural Network (NN)-based prediction model. Next are presented the pseudo-code versions of these two algorithms.

The Random Forest (RF) prediction model is adapted after the one which is described in [41]. The first step of the algorithm is represented by the extraction of a sample S_i that has the size n from S. Using the extracted sample, a tree is fitted using binary recursive partitioning and then all observations are used for initializing a single node. While the stopping criterion is not met, a set of steps are repeated recursively. The prediction for a new point x is performed using the following equation:

$$\widehat{f}(x) = \text{argmax}_y \sum_{j=1}^{J} I\left(\widehat{h}_j(x) = y\right) \tag{2.6}$$

where $\widehat{h}_j(x)$ is the prediction at x using the jth tree.

Algorithm 1: Random Forest Algorithm for Detecting Faulty Products

---

Data: $S = \{(x_1, y_1), \dots, (x_n, y_n)\}$ (training dataset) with $x_i = (x_{i,1}, \dots, x_{i,p})^T$

Result: F_M (a classification model)

1 **for** $j \in \{1, \dots, J\}$ **do**
2   take a bootstrap sample;
3   using the taken bootstrap sample as training data, fit a tree;
4   initially all observations are in a single node;
5   **while** stopping criterion not met **do**
6   randomly select m predictors from the set of available predictors;
7   determine the best binary split;
8   split the node in two nodes using the split from the previous step;
9   **end**
10 **end**
11 return F_M;

The Neural Network (NN) prediction model is adapted after the one that is presented in [42]. Initially, the thresholds and the weights have random values and then the following four steps are repeated M times, where M is the number of iterations: (1) the activation of the network step, (2) the calculation of the outputs of the neurons step, (3) the weight training step and (4) the propagation of the backward error step.

Algorithm 2: Neural Network Algorithm for Detecting Faulty Products

---

Data: $S = \{(x_1, y_1), \dots, (x_n, y_n)\}$ (training dataset), M (number of iterations), H (number of hidden layers)

Result: F_M (a classification model)

1 initialize the weights;
2 $i = 0$;
3 **for** $m \in \{1, \dots, M\}$ **do**
4   activate the network;
5   calculate the outputs of the neurons;
6   update the weights;
7   propagate the errors;
8 **end**
9 return F_M;

2.3.2.8   Bio-Inspired Optimization Phase

In this subsection, there are presented the pseudo-code versions of the algorithms that are used for the selection of the features and for the tuning of the parameters of a deep neural network. These algorithms are inspired from the algorithm that is presented in [43]. Algorithm 3 presents the approach used in this chapter for feature selection using an adapted version of Particle Swarm Optimization (PSO).

Algorithm 3: Particle Swarm Optimization Algorithm for Features Selection

---

Data: $V_{min}, V_{max}, D, w_{min}, w_{max}, c_1, c_2, N_{iterations}, N_{particles}$

Result: g_{best}

1   initialize the swarm: $Swarm = \{p_1, ..., p_{N_{particles}}\}$;

2   **for** i **in** 1 **to** $N_{iterations}$ **do**

3   **for** $p \in Swarm$ **do**

4       $p_{fitness} = ComputeAccuracyOfPredictionModel(p)$;

5       $p_{best} = UpdateLocalBest(p, p_{best})$;

6       $g_{best} = UpdateGlobalBest(p_{best}, g_{best})$;

7   **end**

8      $UpdateInertia(w_{min}, w_{max}, w)$;

9   **for** $p \in Swarm$ **do**

10   **for** $d \in D$ **do**

11        $v_d^{new} = v_d^{old} \times w + r_1 \times c_1 \times (p_{best,d} - x_d^{old}) + r_2 \times c_2 \times (g_{best,d} - x_d^{old})$;

12   **if** $v_d^{new} < V_{min}$ **or** $v_d^{new} > V_{max}$ **then**

13         $v_d^{new} = max(V_{min}, min(v_d^{new}, V_{max}))$;

14   **end**

15        $S(v_d^{new}) = \frac{1}{1+e^{-v_d^{new}}}$;

16   **if** $r_0 \geq S(v_d^{new})$ **then**

17         $x_d^{new} = 0$;

18   **else**

19         $x_d^{new} = 1$;

20   **end**

21   **end**

22   **end**

23   **end**

24   return g_{best};

Algorithm 4 presents the approach used for tuning the number of layers of a deep neural network using an adapted version of Particle Swarm Optimization (PSO).

Algorithm 4: Particle Swarm Optimization Algorithm for Tuning the Layers of a Deep Neural Network

Data: V_{min}, V_{max}, D, w_{min}, w_{max}, c_1, c_2, $N_{iterations}$, $N_{particles}$, $numberOfHiddenLayers$, $neuronsNumberRange$

Result: g_{best}

1   initialize the swarm considering the $numberOfHiddenLayers$ and $neuronsNumberRange$: $Swarm = \{p_1, …, p_{N_{particles}}\}$;

2   **for** i in 1 to $N_{iterations}$ **do**

3   **for** $p \in Swarm$ **do**

4     $p_{fitness} = ComputeDeepNeuralNetworkAccuracy(p)$;

5     $p_{best} = UpdateLocalBest(p, p_{best})$;

6     $g_{best} = UpdateGlobalBest(p_{best}, g_{best})$;

7   **end**

8    $UpdateInertia(w_{min}, w_{max}, w)$;

9   **for** $p \in Swarm$ **do**

10   **for** $d \in D$ **do**

11     $v_d^{new} = v_d^{old} \times w + r_1 \times c_1 \times \left(p_{best,d} - x_d^{old}\right) + r_2 \times c_2 \times \left(g_{best,d} - x_d^{old}\right)$;

12   **if** $v_d^{new} < V_{min}$ **or** $v_d^{new} > V_{max}$ **then**

13      $v_d^{new} = max\left(V_{min}, min(v_d^{new}, V_{max})\right)$;

14   **end**

15   **end**

16   **end**

17 **end**

18 return g_{best};

2.3.2.9   Final Prediction Model Creation Phase

In this phase, the prediction model that gives the best results is chosen. In the case of the machine learning approach, the best prediction model corresponds to the particle associated with the best combination of features and in the case of the deep learning approach, the best prediction model corresponds to the particle associated with the best configuration of a number of nodes of each layer of a deep neural network. The prediction model is evaluated using the testing data that results in the cross-validation phase.

2.3.2.10   Model Evaluation Phase

The metrics that are used for the evaluation of the performance of the machine learning methodology presented in this article are the classical ones and are derived from the elements that compose the confusion matrix [44]:

$$Accuracy = \frac{TP + TN}{TP + TN + FP + FN} \tag{2.7}$$

$$Precision = \frac{TP}{TP + FP} \tag{2.8}$$

$$Recall = \frac{TP}{TP + FN} \tag{2.9}$$

$$F1Score = \frac{2 \times Precision \times Recall}{Precision + Recall} \tag{2.10}$$

where TP is the number of true positives, TN is the number of true negatives, FP is the number of false positives and FN is the number of false negatives.

2.3.3 Data Preprocessing in KNIME (Konstanz Information Miner)

Figure 2.4 presents the data preprocessing steps that are applied in order to transform the data used in experiments in input data for the machine learning algorithms that are applied in order to classify the products as faulty products or not faulty products. The **Excel Reader (XLS)** node is used for reading the input data from an Excel file, the **Missing Value** node replaces the values that are missing using the mean heuristic, the **Normalizer** node normalizes the values, so that they are in the interval [0, 1], the

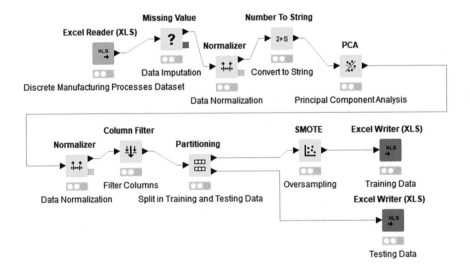

Fig. 2.4 Data preprocessing in KNIME

Number to String node converts the data type of the labels from number to string, the **PCA** node reduces the number of features to a number that is specified as threshold, the **Normalizer** node normalizes the values that result after the transformations from the **PCA** node, the **Column Filter** node filters the initial features that characterize the samples, the **Partitioning** node divides the data into two parts, namely, training data (80%) and testing data (20%) and the **SMOTE** node is used in order to oversample the data.

2.3.4 Discrete Manufacturing Processes Optimization Based on Big Data Technologies

In Fig. 2.5, it is presented an approach for discrete manufacturing processes optimization that is based on big data and semantic web technologies. The data generated by the monitoring sensors are simulated using a Java utility program and is persisted in a Cassandra database. In the approach presented in this chapter, two machine learning approaches are applied: (1) the first approach is based on a Random Forest (RF) prediction model in which the features of the samples that are given as input are selected using an adapted version of the Particle Swarm Optimization (PSO) algorithm for feature selection and (2) the second approach is based on an Artificial Neural Network (ANN) prediction model in which the number of nodes of each hidden layer is determined using an adapted version of the Particle Swarm Optimization (PSO) algorithm for tuning the number of layers of a deep neural network.

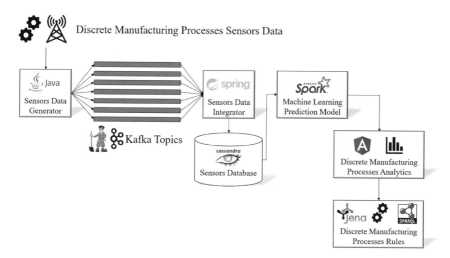

Fig. 2.5 Discrete manufacturing processes optimization conceptual architecture

Table 2.1 Description of experimental datasets after data processing in KNIME	Experimental dataset	Number of features	Number of samples	Number of faulty products
	SECOM dataset	20	13,783	957
	SETFI dataset	20	35,200	17,996
	FL regulators dataset	20	8800	1683

2.4   Results

In this section, there are presented the main experimental results.

2.4.1   Description of the Datasets Used in Experiments

Table 2.1 presents a description of the datasets that are used in experiments after data processing in KNIME (Konstanz Information Miner) [45].

2.4.2   Classification Results

In Fig. 2.6, there are presented the classification results before the bio-inspired optimization phase when the products are classified using RF and ANN.

In Fig. 2.7, there are presented the results after the application of Particle Swarm Optimization (PSO) algorithm for feature selection and for the tuning of the number

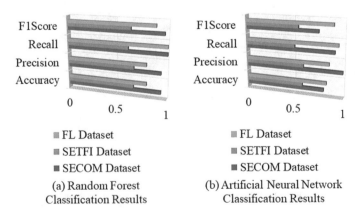

Fig. 2.6 Classification results before bio-inspired optimization phase

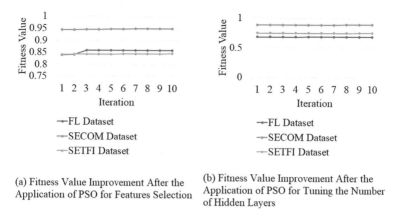

(a) Fitness Value Improvement After the Application of PSO for Features Selection

(b) Fitness Value Improvement After the Application of PSO for Tuning the Number of Hidden Layers

Fig. 2.7 Fitness value improvement after the application of PSO algorithm

of layers of an artificial neural network. In both cases, the algorithms were run for 10 iterations and the number of particles from the swarm was 20.

In Fig. 2.8, there are presented the classification results after the bio-inspired optimization phase when the products are classified using RF and ANN. In the case of the RF classification model, the features are selected using the adapted version of the PSO algorithm for feature selection, and in the case of the ANN classification model, the parameters are tuned using an adapted version of PSO for tuning the number of layers of a deep neural network.

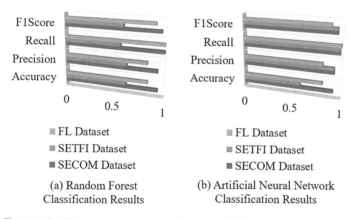

(a) Random Forest Classification Results

(b) Artificial Neural Network Classification Results

Fig. 2.8 Fitness value improvement after the application of PSO algorithm

```
@prefix : <http://www.semanticweb.org/ontologies/manufacturing#> .
@prefix rdf: <http://www.w3.org/1999/02/22-rdf-syntax-ns#> .

[IsFaultyRule:
  (?x rdf:type :Product),
  (?y rdf:type :Features),
  (?x :hasFeatures ?y),
  (?y :hasValue1 ?v1),
  ge(?v1, 0.5),
  (?y :hasValue5 ?v5),
  lessThan(?v5, 0.3)
->
  (?x :hasType :Faulty)
]
```

Fig. 2.9 Illustrative example of Jena rule for detecting if a product is faulty

```
PREFIX rdf: <http://www.w3.org/1999/02/22-rdf-syntax-ns#>
PREFIX manufacturing: <http://www.semanticweb.org/ontologies/manufacturing#>

SELECT ?product WHERE {
  ?product manufacturing:hasFeatures ?features .
  ?features manufacturing:hasValue1 ?v1 .
  ?features manufacturing:hasValue5 ?v5 .
  FILTER (?v1 >= 0.5 && ?v5 < 0.3)
}
```

Fig. 2.10 Illustrative example of SPARQL query for detection of faulty products

2.5 Discussion

In this section, future research directions are presented. In Fig. 2.9, it is presented an illustrative example of a Jena rule [46] for detecting whether the products resulted after the manufacturing processes are faulty.

In Fig. 2.10, it is presented an illustrative example of an SPARQL rule [47] that queries the products that are faulty. The SPARQL rule is derived from the Jena rule that was presented before.

2.6 Conclusions

In this chapter, the digitalization, analysis and optimization of discrete manufacturing processes were approached using Big Data and Internet of Things technologies. The main contributions are: (1) the development of an experimental platform for the analysis of the manufacturing processes in the context of the project OptiPlan, (2) the identification of the products that have faults using a Random Forest (RF)-based approach and an Artificial Neural Network (ANN)-based approach, (3) the optimization of the proposed algorithms using adaptations of the Particle Swarm Optimization

(PSO) algorithm, one adaptation for the selection of features and the other adaptation for the selection of the number of nodes of each layer of a deep neural network and (4) the proposal as future research directions the identification of the faulty products using semantic web technologies based on Apache Jena inference engine and SPARQL query language.

Acknowledgements This work was supported by a grant of the Romanian National Authority for Scientific Research and Innovation, CNCS/CCCDI UEFISCDI, project number PN-III-P2-2.1-BG-2016-0080, within PNCDI III.

References

1. OptiPlan. http://coned.utcluj.ro/OptiPlan/
2. Drath R, Horch A (2014) Industrie 4.0: Hit or hype? [industry forum]. IEEE Ind Eletron Mag 8:56–58
3. Lin Y-C, Hung M-H, Huang H-C, Chen C-C, Yang H-C, Hsieh Y-S, Cheng F-T (2017) Development of advanced manufacturing cloud of things (AMCoT)—a smart manufacturing platform. IEEE Robot Autom Lett 2:1809–1816
4. Lin YC, Hung MH, Wei CF, Cheng FT (2015) Development of a novel cloud-based multitenant model creation scheme for machine tools. In: Proceedings of the 2015 IEEE international conference on automation science and engineering (CASE). Gothenburg, Sweden, pp 1448–1449
5. Behnke D, Muller M, Bok PB, Bonnet J (2018) Intelligent network services enabling industrial IoT systems for flexible smart manufacturing. In: Proceedings of the 2018 14th international conference on wireless and mobile computing, networking and communications (WiMob). Limassol, Cyprus, pp 1–4
6. Lee C, Park L, Cho S (2018) Light-weight Stackelberg game theoretic demand response scheme for massive smart manufacturing systems. IEEE Access 6:23316–23324
7. Kumaravel BT, Bhattacharyya R, Siegel J, Sarma SE, Arunachalam N (2017) Development of an internet of things enabled manufacturing system for tool wear characterization. In: Proceedings of the 2017 IEEE 3rd international symposium in robotics and manufacturing automation (ROMA). Kuala Lumpur, Malaysia, pp 1–6
8. Kim J, Han Y, Lee J (2016) Data imbalance problem solving for smote based oversampling: study on fault detection prediction model in semiconductor manufacturing process. Adv Sci Technol Lett 133:79–84
9. Munirathinam S, Ramadoss B (2016) Predictive models for equipment fault detection in the semiconductor manufacturing process. IJET 8:273–285
10. Dheeru D, Karra Taniskidou E. UCI machine learning repository. http://archive.ics.uci.edu/ml
11. Kerdprasop K, Kerdprasop N (2014) Tool sequence analysis and performance prediction in the wafer fabrication process. Int J Syst Appl Eng Dev 8:268–276
12. AA&YA, Intel. Manufacturing data: semiconductor tool fault isolation. http://www.causality.inf.ethz.ch/repository.php?id=19
13. Zhang J, Wang J, Qin W, Rosa JLG (2016) Artificial neural networks in production scheduling and yield prediction of semiconductor wafer fabrication system. In: Proceedings of artificial neural networks. IntechOpen, London, UK, pp 355–387
14. Mohammadi P, Wang ZJ (2016) Machine learning for quality prediction in abrasion-resistant material manufacturing process. In: Proceedings of the 2016 IEEE Canadian conference on electrical and computer engineering (CCECE). Vancouver, BC, Canada, pp 1–4

15. Sipos E, Ivanciu LN (2017) Failure analysis and prediction using neural networks in the chip manufacturing process. In: Proceedings of the 2017 40th international spring seminar on electronics technology (ISSE). Sofia, Bulgaria, pp 1–4

16. Lee T, Kim CO (2015) Statistical comparison of fault detection models for semiconductor manufacturing processes. IEEE Trans Semiconduct Manuf M 28:80–91

17. Xu S, Li X, Lu WF (2016) Randomized K-d tree ReliefF algorithm for feature selection in handling high dimensional process parameter data. In: Proceedings of the 2016 IEEE 21st international conference on emerging technologies and factory automation (ETFA). Berlin, Germany, pp 1–8

18. Kinghorst J, Geramifard O, Luo M, Chan HL, Yong K, Folmer J, Zou M, Vogel Heuser B (2017) Hidden markov model-based predictive maintenance in semiconductor manufacturing: a genetic algorithm approach. In: Proceedings of the 2017 13th IEEE conference on automation science and engineering (CASE). Xi'An, China, pp 1260–1267

19. Salem M, Taheri S, Yuan J-S (2018) An experimental evaluation of fault diagnosis from imbalanced and incomplete data for smart semiconductor manufacturing. Big Data Cogn Comput 2:1–20

20. Carbery CM, Woods R, Marshall AH (2018) A new data analytics framework emphasising pre-processing in learning AI models for complex manufacturing systems. In: Li K, Fei M, Du D, Yang Z, Yang D (eds) Intelligent computing and internet of things. ICSEE 2018, IMIOT 2018. Communications in computer and information science, vol 924. Springer, Singapore, pp 169–179

21. Wang R, Juang X, Zhao J (2017) Research on multi agent manufacturing process optimization method based on QPSO. In: Proceedings of the 2017 10th international symposium on computational intelligence and design (ISCID). Hangzhou, pp 103–107

22. Doniavi A, Mileham AR, Newnes LB (1999) Optimising electronic manufacturing processes. In: Proceedings of the twenty fourth IEEE/CPMT international electronics manufacturing technology symposium (Cat. No.99CH36330). Austin, TX, USA, pp 63–68

23. Ismail HS, Wang L, Poolton J (2014) A simulation based system for manufacturing process optimisation. In: Proceedings of the 2014 IEEE international conference on industrial engineering and engineering management. Bandar Sunway, Malaysia, pp 963–967

24. Salam B, Gan HY, Lok BK, Albert LCW (2009) Optimizing manufacturing processes of printed electronics. In: Proceedings of the 2009 11th electronics packaging technology conference. Singapore, pp 163–167

25. Hu L, Zhao J, Liang J (2010) Quality rules optimization for manufacturing process based on rough theory. In: Proceedings of the 2010 international conference on measuring technology and mechatronics automation. Chansha City, China, pp 956–959

26. He T (2010) Data mining for the optimization in manufacturing process. In: Proceedings of the 2010 2nd international workshop on intelligent systems and applications. Wuhan, China, pp 1–4

27. Mozumder PK, Loewenstein LM (1992) Method for semiconductor process optimization using functional representations of spatial variations and selectivity. IEEE Trans Compon Hybrids Manuf Technol 15:311–316

28. Kafumbe S (2018) Semiconductor manufacturing process optimisation. In: Proceedings of the 2018 advances in science and engineering technology international conferences (ASET). Abu Dhabi, United Arab Emirates, pp 1–5

29. Qolomany B, Maabreh M, Al Fuqaha A, Gupta A, Banhaddou D (2017) Parameters optimization of deep learning models using particle swarm optimization. In: Proceedings of the 2017 13th international wireless communications and mobile computing conference (IWCMC). Valencia, Spain, pp 1285–1290

30. Desell T, Clachar S, Higgins J, Wild B (2015) Evolving deep recurrent neural networks using ant colony optimization. In: Ochoa G, Chicano F (eds) Evolutionary computation in combinatorial optimization. EvoCOP 2015. Lecture notes in computer science, vol 9026. Springer, Cham, pp 86–98

31. Hafez AI, Zawbaa HM, Emary E, Mahmoud HA, Hassanien AE (2015) An innovative approach for feature selection based on chicken swarm optimization. In: Proceedings of the 2015 7th international conference of soft computing and pattern recognition (SoCPaR). Fukuoka, Japan, pp 19–24
32. Alihodzic A, Tuba E, Tuba M (2017) An upgraded bat algorithm for tuning extreme learning machines for data classification. In: Proceedings of the genetic and evolutionary computation conference companion. Berlin, Germany, pp 125–126
33. Beheshti Z, Shamsuddin SM, Beheshti E, Yuhaniz SS (2014) Enhancement of artificial neural network learning using centripetal accelerated particle swarm optimization for medical diseases diagnosis. Soft Comput Fusion Found Methodol Appl 18:2253–2270
34. Hasan S, Quo TS, Shamsuddin SM, Sallehuddin R (2011) Artificial neural network learning enhancement using artificial fish swarm algorithm. In: Proceedings of the 3rd international conference on computing and informatics (ICOCI). Bandung, Indonesia, pp 117–122
35. Lin KC, Zhang KY, Hung JC (2014) Feature selection of support vector machine based on harmonious cat swarm optimization. In: Proceedings of the 2014 7th international conference on Ubi-media computing and workshops. Ulaanbaatar, Mongolia, pp 205–208
36. Moldovan D, Cioara T, Anghel I, Salomie I (2017) Machine learning for sensor-based manufacturing processes. In: Proceedings of the 2017 13th IEEE international conference on intelligent computer communication and processing (ICCP). Cluj-Napoca, Cluj, Romania, pp 147–154
37. Moldovan D, Chifu V, Pop C, Cioara T, Anghel I, Salomie I (2018) Chicken swarm optimization and deep learning for manufacturing processes. In: Proceedings of the 2018 17th RoEduNet conference: networking in education and research (RoEduNet). Cluj-Napoca, Cluj, Romania, pp 1–6
38. Anghel I, Cioara T, Moldovan D, Salomie I, Tomus MM (2018) Prediction of manufacturing processes errors: gradient boosted trees versus deep neural networks. In: Proceedings of the 2018 IEEE 16th international conference on embedded and ubiquitous computing (EUC). Bucuresti, Romania, pp 29–36
39. Silipo R, Adae I, Hart A, Berthold M (2015) Seven techniques for dimensionality reduction missing values, low variance filter, high correlation filter, PCA, random forests, backward feature elimination, and forward feature construction, pp 1–21
40. Chawla NV, Bowyer KW, Hall LO (2002) SMOTE: synthetic minority over-sampling technique. J Artif Intell Res 16:321–357
41. Cutler A, Cutler DR, Stevens JR (2001) Random forests. Mach Learn 45:1–20
42. Bhuiyan A-A, Liu CH (2007) On face recognition using Gabor filters. World Acad Sci Eng Technol 28:51–56
43. Kennedy J, Eberhart RC (1997) A discrete binary version of the particle swarm algorithm. In: Proceedings of the 1997 IEEE international conference on systems, man, and cybernetics, computational cybernetics and simulation. Orlando, FL, USA, pp 4104–4108
44. Confusion matrix. http://en.wikipedia.org/wiki/Confusion_matrix
45. Berthold MR, Cebron N, Dill F, Gabriel TR, Kotter T, Meinl T, Ohl P, Sieb C, Thiel K, Wiswedel B (2008) KNIME: The konstanz infomation miner. In: Preisach C, Burkhardt H, Schmidt Thieme L, Decker R (eds) Data analysis, machine learning and applications. Studies in classification, data analysis, and knowledge organization. Springer, Berlin, Heidelberg, pp 319–326
46. Apache Jena. http://jena.apache.org/documentation/inference/
47. Prashanth Kumar KN, Ravi Kumar V, Raghuveer K (2017) A survey on semantic web technologies for the internet of things. In: Proceedings of the 2017 international conference on current trends in computer, electrical, electronics and communication (ICCTCEEC). Mysore, India, pp 316–322

Chapter 3
An Empirical Study on Teleworking Among Slovakia's Office-Based Academics

Michal Beno

Abstract Teleworking has become fairly popular in recent years, mostly outside the academic setting. The evolution of the information age offers the flexibility to perform work-related tasks anywhere at any time. Over the past several years, changes in the workplace within higher education have led to the notion of teleworking, making it possible for employees to work from a remote location. However, very little research has been done on teleworking practices within the higher-education environment. Generally, the concept of teleworking or telecommuting is less common in Slovakia than in other countries. The introduction of teleworking in many different organisations, including universities, could provide great advantages. The success rate of the implementation of teleworking in universities is very high because of the great number of researchers. A study was conducted to explore the potential of introducing teleworking at universities in Slovakia. Drawing on the existing literature related to teleworking and the outcome of an empirical study, this paper presents enlightening findings on teleworking at these organisations in Slovakia. The paper investigates the attitudes and viewpoints of potential teleworkers toward the possibility of introducing teleworking in universities. The research reveals high levels of support for teleworking practices, which are associated with high levels of work productivity and satisfaction (student satisfaction), lower levels of emotional and physical fatigue, reduced work stress, absence, frustration and overload. Overall, this critique presents insightful findings on telecommuting practices within an academic setting, which clearly signal the potential for a shift in the office culture at higher-education institutions that provide distance tuition. The study makes a significant contribution to a limited collection of empirical research on telecommuting practices at universities and also guides institutions in refining and/or redefining future teleworking strategies or programmes.

Keywords Telecommuting and teleworking · Academia · Slovakia

M. Beno (✉)
The Institute of Technology and Business in České Budějovice, České Budějovice 370 01, Czech Republic
e-mail: beno@mail.vstecb.cz

© Springer Nature Switzerland AG 2021
F. Pop and G. Neagu (eds.), *Big Data Platforms and Applications*,
Computer Communications and Networks,
https://doi.org/10.1007/978-3-030-38836-2_3

3.1 Introduction

The industrial revolution brought employees from their homes to the factories. With information and communication technology (ICT), it is possible for employees to return to their homes again. The idea of working for a central organisation from home or from one's own environment emerged in the 70s [1]. Technology has had a significant impact on work, making it possible to work from the home and other locations [2], i.e. work can be done anywhere at any time [3]. The information age, where ICT plays a very important and vital role, allows people to perform their daily work both inside and outside the office premises using ICT tools. According to Bredin [4], the development of the Internet, the low price of fast computers, and the developments of videoconferencing, groupware, digital phones and satellite communications have made teleworking and virtual offices more feasible and popular. Today's office work can be carried out from practically any location and at any time. Nowadays, the practice of various flexible work schedules has a growing number of supporters, taking into account that, among other things, it reduces the cost of office space, overcomes geographical borders and the limitations of hiring the best quality workers and provides a better work/life balance for employees. Wherever labour laws are flexible, this brings benefits for both companies and their employees [5]. There are many advantages, especially flexibility, easier conditions, productivity, improved communications and saving of the time spent that would be spent commuting.

Teleworking, as commonly used in Europe, or telecommuting in the USA, is in our opinion a work phenomenon that allows human beings to work outside a localised workplace rather than inside it, using ICT equipment. The term telecommuting first appeared in the USA in the 70s [6]. Currently, it is used worldwide in different concepts such as telework, e-Work, online work, nomadic work, mobile work and by various professions such as programmers, technicians, customer service staff, marketing consultants, purchasing agents, journalists, market investigators, publishers, interpreters, brokers, accountants and many others.

There is not much existing literature on research about academic teleworkers, and what is available is limited to a few international examples, e.g. collective work published by the Athabasca University on the potential benefits and drawbacks for telecommuting academics and the opportunities and challenges facing their institution [7].

In the main European countries, the innovative forms of work organisation are already relatively well known and widespread [8, 9]. According to the Sixth European Working Conditions Survey [10], most workers (62% of men and 78% of women) have a single main place of work almost all the time. Nearly a third of workers (30%) divide their working time across multiple locations. Despite the popular image of mobile workers as young knowledge workers typing away on their laptops in a park or a café, it is more common in the construction (57%), transport (49%) and agriculture (50%) sectors to have more than one regular place of work.

In Slovakia, teleworking is expanding very slowly [11]. We think this is probably a result of conservatism among employers and employees, as well as because of a lack of information about this form of work among the population as a whole. The legislation of most European countries provides that the employment contract must contain a precise definition of the place of work (e.g. France, Austria, Hungary). In some countries, an employee who works at/from home or at other premises has the same rights as an office employee (Spain, Portugal, the Netherlands). Telework in Slovakia was implemented by an amendment of the Labour Code (Act No. 348/2007, effective as from 1 September 2007) to bring Slovakian labour legislation in line with EU legislation in respect of new elements, such as homework and telework (Section 52) [12]. In Slovakia, an employee is not considered to be performing homework or telework if he/she works at home or at another agreed workplace only occasionally or in exceptional circumstances with the consent of the employer or in terms of an agreement with him/her, subject to the condition that the type of work that the employee performs is allowed under the terms of the employment contract [13].

This paper will define teleworking, look into the uptake of teleworking in Slovakia and examine the advantages and disadvantages of this form of work. Drawing on a review of the literature on the research and practices of telecommuting as an alternative work arrangement outside the academic sphere, this paper discusses some of the major technological issues, the potential benefits and drawbacks of telecommuting for telecommuting academics, and the potential opportunities for and challenges faced by their institutions. Though critically important, a detailed examination of the technologies and the role that information and communication technologies play in enhancing the effectiveness and efficiency of a dispersed faculty work environment is beyond the scope of this paper. The issues discussed in this paper are confined to those relevant to full-time academics teaching in education institutions as core faculty. This paper discusses some of the issues involved in introducing teleworking in Slovak academic society. Finally, the paper closes with the conclusions drawn from the investigation.

3.2 Methodology

As more and more people have come online, qualitative researchers have explored the use of online tools for research [14]. These tools include e-mail interviewing, instant messaging, Skype [15], and (a)synchronous interviewing [14, 16].

It was found that e-mail interviewing offers unprecedented opportunities for qualitative research and provides access to millions of potential research participants who would otherwise be inaccessible. The method can be employed quickly, conveniently and inexpensively and can generate high-quality data when used with due care. The method involves multiple e-mail exchanges between the interviewer and interviewee [14]. This means e-mail interviews can be conducted with participants all over the world without the additional expense of travel costs and travel time. It makes it possible to conduct asynchronous interviews. According to Walker [17], a major advantage of the e-mail interview is that it offers a convenient and practical alternative to overcoming geographical barriers and financial concerns that hinder face-to-face interviews. Furthermore, qualitative researchers using e-mail interviews for data collection found that the scheduling convenience of e-mail interviews increases access to participants and encourages greater participation by working adults [18]. A disadvantage of this interviewing tool for participants is that drafting written responses is by nature more time consuming than oral interviews [18]. In our view, the lack of e-mail access or discomfort with e-mail communication may generally limit participation.

This study was undertaken to ascertain the perceptions of the employees of Slovakian universities (private and state) toward participating in teleworking as an alternative work arrangement.

E-mail interviews with questions were prepared and distributed to both academic private and state universities in Slovakia via e-mail. The e-mail interview was drawn up in Word and consisted of two sections with six open questions. The first (control) section included the gender, age, job position, marital status and computer ownership characteristics of the interviewee. The second section consisted of perceptions of teleworking and dealt with the willingness to do telework, frequency of teleworking, reasons for teleworking and reasons for not liking teleworking, job tasks performed offsite, ICT equipment, and impact on individuals and the organisation. E-mail interviews were particularly well suited to our study as they are a very good way of producing detailed written accounts of the participants' experiences and memories relating to long-term involvement in the teleworking field.

The study targeted academics who are office-based (non-teleworkers). The primary aim of our study was to establish the attitudes and perceptions of onsite workers regarding teleworking and to determine the impact and effect (emotions/attitudes) of teleworking on office-based workers.

The sample size of the interviewees who participated in the teleworker evaluation study and whose interviews were included in this research was 125 academics. Some may believe that e-mail interviewing is more suited to research with young participants, but we have found e-mail interviews to be popular with respondents in a wide range of age groups. Furthermore, the answerers needed to feel comfortable writing the accounts of their experiences. The data were received via e-mail in the period 1 February to 31 March 2018 and were then edited and prepared for data analysis and interpretation.

3.3   Meaning of Telecommuting or Teleworking

Besides the original explanation of telecommuting by Nilles [19], the literature review shows many common conceptions of this phenomenon. Taking this into account, the following chosen definitions provide a broader understanding of this concept:

(a)   "the utilisation of telecommunication and computer technologies to replace or reduce traditional commuting to the workplace, and telecommuters most often work in management and service jobs where there is little need for physical interaction with goods or people in the workplace", and Marcus [20] emphasised that "quantifying telecommuters is difficult for a number of reasons";

(b)   "work arrangement in which employees perform their regular work at a site other than the ordinary workplace, supported by technological connections" [21];

(c)   "periodic work out of the principal office, one or more days per week either at home, a client's site, or in a telework centre" [22];

(d)   working outside the conventional workplace and communicating by means of communications or computer-based technology [3];

(e)   "a subset of alternative work options where work is conducted at an off-site location and the employee uses telecommunications technology, including computers, video, and telephone systems, fax machines and high-speed hook-ups for data transfers" [23];

(f)   when the concept of teleworking emerged about 20 years ago, it referred to the practice of working from home using telecommunication links to replace commuting, and was sometimes known as telecommuting [24];

(g)   "telework is a form of organising and/or performing work, using information technology, in the context of an employment contract/relationship, where work, which could also be performed at the employer's premises, is carried out away from those premises on a regular basis" [25];

(h)   "the term 'telework' or 'teleworking' refers to a work flexibility arrangement under which an employee performs the duties and responsibilities of such employee's position, and other authorised activities, from an approved worksite other than the location from which the employee would otherwise work" [26].

Telework, also known as telecommuting, can be defined simply as when employees work at some place other than the traditional workplace [27]. The concept of telework, more precisely telecommuting, was born during the oil crisis in the early 1970s, when American Jack Nilles and colleagues published their calculations on the savings to the national economy that would result from reduced commuting [6]. Currently, it is used worldwide in different concepts such as telework, e-Work, online work, nomadic work, mobile work and by various professions such as programmers, technicians, customer service staff, marketing consultants, purchasing agents, journalists, market investigators, publishers, interpreters, brokers and accountants [28]. On the basis of Chaudron [29], it can be stated that teleworking can be successfully implemented when the right reason, the right job, the right employee, the right manager and the right environment are all present.

The main kinds of telework can be further distinguished into several forms, such as home-based, satellite offices, neighbourhood work centres and mobile work [3, 30–32]. A more detailed description of the four main forms of de-localisation of work follows:

Home-based telecommuting is usually performed in a dedicated area of the worker's place of residence, as illustrated in Fig. 3.1. For this type of telecommuting, equipment installation fees are usually subsidised, entirely or in part, by the organisation [33–35]. Organisationally, telecommuting is part of the main workplace. It should be voluntary. Home-based includes different work types such as running as one's only job; running two various jobs; bringing work home (overtime) and traditional form of telecommuting. This is described by the term small office/home office (SOHO) work done at home:

Satellite offices are geographically removed from the central office but perform similar types of work as the central office (see Fig. 3.2). According to Di Martino and

Fig. 3.1 Home-based telecommuting (diagram by author)

Fig. 3.2 Satellite offices (diagram by author)

Fig. 3.3 Mobile workers [42]

Wirth [36], a satellite work centre is a separate unit within an enterprise, geographically removed from the central location but remaining in constant communication with it. These offices take the form of small organisational affiliates, generally located close to residential areas, where a telecommunications link with the head office is permanently maintained [35, 37] and can be classified into two types, traditional and non-traditional [38]. Generally, workers perform their duties in scattered small units of a central office;

Neighbourhood Work Centres (often called telecottages) or telework centres are similar to satellite offices but involve multiple organisations, and employees from different organisations, or self-employed entrepreneurs, share space and equipment in the work centre that is closest to their home [36, 39];

Mobile work (outside the home or main office) is similar to telecommuting, but workers stay at the customer's site or front-line and work from there rather than at home. Mobile workers receive work orders and deliver work results by way of computers and communications technology, as shown in Fig. 3.3. According to the Cisco Study [40], mobile working can improve productivity, motivation, flexibility and staff retention. The definition of mobile work is not simple. On the basis of electronic commerce and telework trends, mobile workers are those who work at least 10 h per week away from home and the main place of work, e.g. on business trips, in the field, travelling or at the customer's premises [41].

Work from home can be carried out only in terms of a standard contractual employment relationship between employer and employee. This means that the rights and obligations of the employer and employee are set out in the law. Work from home is always subject to internal entrepreneurial agreements. This means that an employer cannot force an employee to work from home, and an employee cannot demand that an employer must allow it. If an employee works from home, the number of working hours is regulated by the contracted job time, as in any other type of job. However, employees working from home can schedule their working time to suit their particular circumstances. The employees are expected to decide for themselves when to work (schedule flexibility), where to work (telecommuting), and by means of which communication tool/medium (smartphone, e-mail, videoconference) to work [43, 44].

Table 3.1 Advantages and disadvantages of telework in both sectors

Advantages	Disadvantages
Greater autonomy [7]	Increased family/work conflicts [7, 54]
Better personal family and work/life balance [7, 45, 46]	Lower perceived personal growth and career advancement [7, 54]
Comfortable home working environment [7]	Less interaction with colleagues and lower sense of belonging [7]
Opportunity to take advantage of the geographic location to conduct research and provide services to communities [7]	Inadequate work environment at home [7]
Reduction/elimination of transport time [35, 47–49]	Occupational and health issues [2, 7]
Cost savings related to work habits (e.g., clothing and food) [47, 48]	Costs of running the home office [7]
Flexibility in the organisation of working hours and leisure activities [36, 49–51]	Feeling of isolation [7, 50, 55]
	Perceived as undervalued [56]
Increase in productivity [2, 7, 36, 48, 50, 52, 53]	Overwork [7, 21, 35]
Increase in productivity and creativity [49]	Decrease in frequency of intra-organisational communication [57, 58]

Telework potentially impacts individuals in various ways. We have, therefore, provided a summary in Table 3.1 of the advantages and disadvantages of telework in the public and private sectors.

3.3.1 Teleworking in Slovakia

As stated in the introduction, Slovakia legalised the inclusion of telework in a work contract in 2007 (Labour Code, Section 52). Section 52 of the Labour Code defines telework as "a labour relation of an employee who performs work for an employer at home or at another agreed place, pursuant to conditions agreed in the employment contract, using information technology (hereinafter referred to as 'telework') within the working time arranged by himself/herself," and states that telework "shall be governed by this Act" [59].

Since then, we have observed the spread of this work form mostly in the private sector in different professional fields, such as IT, as well as among marketing consultants, purchasing agents, brokers, journalists, financial analysts and accountants. In the public sector in this country, telework is not very widespread.

The use of telework is supported in several documents and strategies, e.g. the National Employment Strategy of the Slovak Republic, and the National Reform Programme of the Slovak Republic (especially the fields of employment and social inclusion). At the local level, there are also initiatives by various NGOs that encourage telework by their employers and employees. The Statistical Office of the Republic of Slovakia does not provide data on the use of telework. Nevertheless, telework is used as a method of work by 3.1% of workers in Slovakia [60]. According to Kordosova [61], the use of telework in Slovakia is low compared with its use in other EU countries, because the information technology (IT) required for telework is only available in a limited number of households, and employees in Slovakia are more interested in traditional forms of work than in greater flexibility.

The latest data from Statista in 2015 show the percentages of employees in different European Union countries who can do telework or perform ICT mobile work, either on a regular basis or occasionally. In Denmark, 37% of workers are able to work from home on occasion, the highest in any country. In Slovakia, the figure is only 10% [62]. In the European Union, only 5% of employees worked from home in 2017. The highest percentage of these workers was in the Netherlands (13.7%), and the lowest in Bulgaria (0.3%). More women (5.3%) than men (4.7%) work from home. But only 1.6% of EU citizens in the age group 15–24 years do home-based work, 4.7% of those from 25 to 49 and 6.4% of those over 50 [63].

Historically, Slovakia's labour market in the twentieth century has been characterised by rigid labour regulation, the prevalence of permanent employment and full-time work and the absence of non-standard kinds of work (such as work from home and work in call centres). Nowadays, we notice different non-standard employment in Slovakia, e.g. temporary employment, part-time work, flexible working time and shift work. According to the OECD data of 2017, Slovakia has 9.6% dependent employment, compared with the average of 11.2% for OECD countries [64]. Less common forms preferred mostly by foreign-owned companies are working from home, teleworking, on-call working, remote working, mobile working and e-working.

As far as we are able to determine, it appears that obstructions to the increase of non-standard kinds of employment do not lie only in legislation, but also in society, wages, working conditions and social security. The Slovak view of non-standard employment is that it is associated with low wages and poor working conditions; it is associated with female work and suspicious circumstances; and it is synonymous with degrading work or with unstable job opportunities.

Hrubesova and Matulnikova [5] further add "rigid legal regulations in connection with uneven working time distribution, and lack of legal regulation in connection with the specific arrangement of home office, do not protect employees nor support business. We believe effective dialogue on a proposal of the new legal regulation of flexible working arrangements would be much appreciated by employers as well as employees".

3.4 Office-Based Teleworking Results

In our study, teleworking is a regularly scheduled flexible work arrangement whereby employees are authorised by the appropriate management to perform the normal duties and responsibilities of their position, through the use of computers and other telecommunications, at designated sites other than their normal place of work.

Teleworking does not apply in cases where employees work at home on an incidental or occasional basis (i.e. to complete regular or special projects that require concentration and fewer interruptions). These alternate work-site arrangements can be approved by an employee's department on a case by case basis without written agreement.

This study presents the final results of e-mail interviews among non-teleworking academics.

To determine the effect of teleworking on office-based staff, we carried out a study in which a total of 125 office-based academics participated. The section on social characteristics included age, gender, marital status and job position. The results revealed that the majority of the respondents (44%) was in the age group 31–40 years, followed by 33.6% aged 20–30 years, 12.8% were 41–50 years and 9.6% were over 50 years of age. In terms of gender, the interviewees were found to be almost equally distributed, with the proportion of males being 52% and females 48%. This is worth noting because there is usually a gender bias in response rates in favour of men. It is possible that the male population is more at ease with the idea of e-mail interviewing compared with women. In general, men are perhaps more comfortable with providing written records of their experiences. In Slovak society, however, there is a clear gender equality. In any event, we are of the opinion that the topic made men more likely to participate than women. The majority were married (71.2%). The full results of these characteristics are shown in Table 3.2.

From the results of the study, we observe, in general, a high potential to introduce teleworking as an alternative work arrangement at universities. This is mainly on account of the nature and function of the work carried out by academics. The primary tasks are obviously teaching, conducting research, consulting and preparing projects. The use of ICT by staff members allows them to perform their work at a distance, e.g. distance teaching, IT networking or using a learning platform such as Moodle. It is evident that there is a high penetration level of personal computers and tablets at home but, as one interviewee remarked, "without a reliable internet connection, telework is impossible". Our findings reveal the importance of reliable Internet and phone connections for teleworkers. In Slovakia, the rate of Internet users increased (4,477,641) and non-users decreased (951,777) in 2016 [65]. Some of the respondents appeared to be at a disadvantage in terms of remaining up-to-date with computer systems.

The success of the implementation of teleworking among academics is very much related to the rate that staff members accept doing telework. Interestingly, the interviewed academics revealed a positive attitude toward the acceptance of this work form as an alternate work arrangement. A total of 82.4% of the respondents indicated they would be interested in adopting telework if given the option by the university authorities. However, the data on the extent of the willingness to do telework revealed that only 36% of the respondents would prefer to do telework on 2 days per week, and 20% on 1 day per week. One academic commented as follows: "I would not want to work from home more days because I come out to work as I want to mix with

Table 3.2 Characteristics of respondents (N = 125)

Characteristics	Number	Percentage
Age		
20–30 years	42	33.6
31–40 years	55	44.0
41–50 years	16	12.8
>50 years	12	9.6
Gender		
Male	65	52.0
Female	60	48.0
Marital status		
Married	89	71.2
Unmarried	36	28.8
Job position		
Professor	8	6.4
Associate Professor	41	32.8
Assistant Professor	17	13.6
Lecturer	30	24.0
Assistant Lecturer	19	23.2

Source Author

people. Sometimes you want to just chat about things with someone and ask them their opinions or just discuss something about a decision as a more natural flow but it depends on the person". Another said he would like "to enjoy Sunday, going to bed later and not having to come to the office early in the morning on Monday because of the traffic … one day of telework is the best solution". The next non-teleworker emphasised that "I would prefer to work from home only two days as I save the time to commute from the office".

According to the results of our study, the practice of working at a distance has benefits such as flexibility (84%), less commuting (76%), increased productivity (70.4%), better concentration (76%), better work/life balance (68%).

"Work flexibility could allow me to manage tasks, workload, organisation, travel better," a non-teleworker said about the work benefits. Our results indicate that this kind of work allows people to manage family responsibilities more easily. "Telework means no interruptions, as often happens in the office, and better follow-up on the work," stated one academic, and added, "I feel more frustrated and overloaded with my work and work performance in the office". Two respondents stressed how much concentration improves: "Better concentration for writing documents, projects, paperwork"; and, "Calmness to concentrate on tasks, leads to better thinking and writing". Our findings indicate that the following characteristics are important qualities for teleworkers: self-discipline (60%), the ability to work alone (44%), being organised (41.6%), ability to solve problems independently (40%), technological literacy

(52%), good communication skills (60%), good relationship with manager or supervisor (54.4%), high motivation (57.6%). An office-based worker revealed that it is important to maintain good relationships with colleagues and managers: "Communication is important ... good communication between colleagues, better cooperation, ability to work in a team, to spread the work out fairly, to avoid isolation and to plan meetings are the main issues".

Isolation (68%), less inducement to do more work (60%), decreased chances of promotion (46.4%), increased confrontation between family members (36%) and technical difficulties (60%) are a few of the disadvantages revealed by the survey data. Non-teleworkers claimed that technical difficulties arose when teleworkers have to solve problems without support from the organisation at a distance: "Difficult to reach colleagues when IT system is down". Although telework allows increased work/life balance, it can also be a source of overworking or burnout.

Isolation is one of the principal factors that put employees off telework because of less socialisation with co-workers in the central office. One respondent claimed that increased independence can be a disadvantage: "Risk of isolation. Risk of losing touch with the day-to-day activities of the department (illness of colleagues, team-work aspects). The necessity to adapt the job, tasks and objectives, knowing that independence could also be a disadvantage"; "Risk of becoming isolated from the team because of more rigorous work, longer working hours". Our findings reveal that it is important for teleworkers to be involved in office life. A possible solution could be a telepresence robot for reducing the feeling of isolation, increasing interaction with colleagues and providing a greater sense of belonging [28].

One non-teleworker emphasised that there is not so much stress in telework because of less office-sharing: "Depending on the circumstances, conflict could arise in an office situation, but this is easier to handle or resolve in an office; however, there would probably be less conflict without the face-to-face contact of colleagues, and therefore less stress". Another one argued that "because of your physical absence, it could lead to you being excluded, having less presence in the team". Non-teleworkers also claimed that they had less awareness of unofficial information: "more difficult communication", "lack of contact with colleagues". Generally, the lack of face-to-face contact between teleworkers and non-teleworkers in the telework context can have a negative effect. A total of 36% of participants felt that for certain tasks, teleworkers should be seen in person. Telepresence is a mix of technologies, which allows users to be present remotely. Widely used telepresence tools are videoconference applications like Appear.in, Skype, Messenger, Google Hangouts and Viber [28], which are useful for overcoming these hurdles.

The absence of fixed working times would allow teleworkers to work at their own pace, and the flexibility would result in more work being done at home in comparison with the central office. The study showed that nearly 17.6% of the respondents were not in favour of telework for one or more reasons.

3.5 Discussion

How people manage the process of going to work, doing their work and then leaving work again was defined by 1870, perfected by 1930, and has changed little since [66]. However, in recent years, telecommuting/teleworking has become more popular because of enhancements of modern technology, changes in the economy and shifting views in our society.

This flexible concept of working anywhere at any time and in any way has been implemented in many organisations, including the academic society. Allowing employees to work from home is becoming a common topic of discussion in human resources circles. One of the main focuses of this discussion is how this trend affects employers, employees, companies, students and society. It might be a surprise to learn that our study has found there is a positive attitude toward the acceptance of this form as an alternate work arrangement, with its accompanying increased productivity. To obtain a better understanding of this trend, we have listed the most important pros and cons relating to this topic.

With the worldwide rise of teleworking within the workplace, it is important to understand this change over time, with its implications for workplace policies. As Markarian [67] emphasises, for the right person, telecommuting works, but who are the right people? What do employees and employers need to know regarding the concept of teleworking, such as the pros, cons and risks already mentioned, as well as implications for the work environment and our society?

Noble [68] states that "university resources were reallocated towards research at the expense of its education function. Class sizes swelled, teaching staffs and instructional resources were reduced, salaries were frozen, and curricular offerings were cut to the bone. At the same time, tuition soared to subsidise the creation and maintenance of the commercial infrastructure (and correspondingly bloated administration), which has never really paid off. In the end students were paying more for their education and getting less, and the campuses were in crisis". We believe that teleworking affords institutions of higher education the opportunity to reduce costs.

3.6 Conclusions

To sum up, changes in technology, economy and society throughout the past century have led to various changes in the workplace and in higher education where jobholders no longer need to travel to the workplace to complete their tasks. As a result, employees are now able to work anywhere at any time and in any way, which has benefits for the employee, the employer and society. In the academic world, staff, students and education as a whole benefit from this kind of flexibility.

This paper presents the results of our survey of the possibility of office-based workers doing academic work at a distance, telework, in Slovakia. The meaning of telework in this study is a regularly scheduled flexible work arrangement whereby the employees are authorised by appropriate management to perform the normal duties and responsibilities of their position, through the use of computers and other means of telecommunication, at designated sites other than their normal place of work.

This study targeted academics who are office-based (non-teleworkers). The primary aim of our study was to establish the attitudes and perceptions of on-site workers regarding teleworking and to determine the impact and effect of teleworking on the emotions and attitudes of office-based workers.

Most of the previous research has been based on the influence of ICT and its importance for telework. Our purpose was to provide an evaluation framework that looks at telework from the point of view of academics situated in the office.

According to Zulova [12], "the Slovak Republic implemented the Telework Agreement through national legislation". Evidently, it is necessary to increase the awareness of telework in Slovakia by different means, including digitisation and modernised regulation of working conditions.

The advancement and proliferation of ICT help to accomplish working tasks more quickly and effectively. This development also increases the penetration level of computer ownership among individuals, mainly because of a decrease in the cost of personal computers. This trend is expected to expand further until every individual owns a personal computer at home. This may eventually also increase the number of teleworkers, either working formally or on an informal basis.

The initial findings showed a generally high level of acceptance among employees, especially academic staff, of adopting teleworking as an alternative work arrangement. One of the main reasons for the high acceptability arises from the fact that the academic staff of the university spend a great deal of time on independent research activities.

This modern kind of work has positive as well as negative aspects. On the positive side, telework allows the work to be carried out anywhere and at any time, which has the obvious advantages of improving flexibility and expanding workforce participation for those who may face difficulties in travelling to the office. Of course, it also saves the time that would otherwise be spent commuting to work. On the negative side, the main disadvantages arise from absence from the office and the subsequent loss of contact with the personal activities of other staff members, as well as, and probably more importantly, the social interaction with colleagues. For example, one cannot help but wonder how many theories in the field of sociology owe their genesis to chance remarks among colleagues in an informal situation.

In summary, this paper outlined some of the key issues related to the implementation of teleworking among office-based academics from a Slovakian perspective. By examining the relevant literature, we illustrated that there are a number of important advantages and disadvantages (equivalent to recent study with significant pros and cons [69]), and finally, we proposed that though teleworking may not be able to solve all the issues involved in co-working, it does offer a promising new way of working that can contribute to individual and organisational flexibility and ultimately to improved productivity.

Regardless of the manner of its use, teleworking affects the environment for academics as a result of the changes in the utilisation of ICT, communication, management and trust. Because of the rising popularity of telework [70], further research on the topic is needed for the remaining issues and to provide explicitly for future practices in the academic field. Companies and organisations supporting the concept of teleworking have to take many challenges into account. Changing one's workplace from a conventional office-based environment to a home-based or a different location necessarily involves changes in mutual cooperation, frequency, quality and interaction between all the parties involved. Work of this nature has the ability to influence this relationship between the participants. We believe that teleworking cannot replace normal work arrangements; it can only be added to them so as to meet the employee's needs and demands.

References

1. Burch S (1991) Teleworking: a strategic guide for management. Kogan Page, London
2. Grant CA, Wallace LM, Spurgeon PC (2013) An exploration of the psychological factors affecting remote e-worker's job effectiveness, well-being and work-life balance. Empl Relat 35(5):527–546
3. Bailey D, Kurland NB (1999) Telework: the advantages and challenges of working here, there, anywhere, and anytime. Organ Dyn 28:53–68
4. Bredin A (1996) The virtual office survival book. Wiley, New York
5. Hrubesova Z, Matulnikova K (2019) amcham.sk. http://www.amcham.sk/publications/connection-magazine/issues/2014-05/126770_flexible-employment-opportunity-or-risk. Last accessed 14 February 2019
6. Nilles JM, Carlson RF, Gay P, Hanneman GJ (1973–1974) The telecommunications-transportation tradeoff. An Interdisciplinary Research Program of University of Southern California, Los Angeles. NSF-RA-5-74-020
7. Ng F (2006) Academics telecommuting in open and distance education universities: issues, challenges, and opportunities. Int Rev Res Open Distance Learn 7(2):16 pp
8. Eurofound, Telework in the European Union. https://www.eurofound.europa.eu/sites/default/files/ef_files/docs/eiro/tn0910050s/tn0910050s.pdf. Last accessed 16 February 2018
9. SIBIS, Statistical Indicators, Benchmarking the Information Society. http://www.sibis-eu.org/about/about.htm. Last accessed 16 February 2018
10. Eurofound, Sixth European Working Conditions Survey (2016). https://www.eurofound.europa.eu/sites/default/files/ef_publication/field_ef_document/ef1634en.pdf. Last accessed 14 February 2019
11. Babos P (2013) Pracovne podmienky v typickych, netypickych a velmi netypickych zamestnaniach na Slovensku: mapovacia studia. SAV, Prognosticke prace, 5, c. 3
12. Zulova J (2017) Implementing telework agreement in Slovakia. ESJ 2017, pp 20–30
13. Employment.gov.sk., Labour Code, Slovak Republic–Slovakia. https://www.employment.gov.sk/files/praca-zamestnanost/vztah-zamestnanca-zamestnavatela/zakonnik-prace/zakonnik-prace-anglicka-verzia-labour-code-full-wording-2012.pdf. Last accessed 14 February 2019
14. Salmons J (2010) Online interviews in real time. Sage, Thousand Oak
15. Lo Iacono V, Symonds P, Brown HKD (2015) Skype as a tool for qualitative research interviews. Sociol Res Online 21(2):12. https://warwick.ac.uk/fac/soc/al/people/mann/interviews/paul_symonds_-_skype-research-method.pdf. Last accessed 19 March 2018
16. Bowden Ch, Galindo-Gonzales S (2015) Interviewing when you're not face-to-face: the use of email interviews in a phenomenological study. Int J Dr Stud 10:79–92
17. Walker D (2013) The internet as a medium for health service research. Part 1. Nurse Res 20(4):18–21

18. Fritz RL, Vandermause R (2017) Data collection via in-depth email interviewing: lessons from the field. Qual Health Res 1–10
19. Nilles JM (1975) Telecommunications and organizational decentralization. IEEE Trans Commun 23:1142–1147
20. Marcus J (1995) The environmental and social impacts of telecommuting and teleactivities. Environmental Studies 196. University of California, Santa Cruz, CA, pp 1–27
21. Fitzer MM (1997) Managing from Afar: performance and rewards in a telecommuting environment. Compens Benefits Rev 14(4):65–73
22. Nilles JM (1998) Managing telework: strategies for managing the virtual workforce. Wiley, New York
23. Kossek E (2003) Telecommuting, a sloan work and family research network encyclopedia entry. Sloan Work and Family Research Network, Chestnut Hill, MA
24. Ruiz Y, Walling A (2005) Home-based working using communication technologies. Labour Market Trends 113:417–426
25. Baltina I (2012) Overview on European policies on TELEWORK. Institute of National and Regional Economy, Riga, pp 1–12
26. Telework Enhancement Act of 2010. https://www.congress.gov/111/plaws/publ292/PLAW-111publ292.pdf. Last accessed 19 March 2018
27. Beno M (2018) Transformation of human labour from stone age to information age. In: Younas M, Awan I, Ghinea G, Catalan Cid M (eds) Mobile web and intelligent information systems. MobiWIS 2018. Lecture notes in computer science, vol 10995. Springer, Cham, pp 205–216
28. Beno M (2018) Work flexibility, telepresence in the office for remote workers: a case study from Austria. In: Multi-disciplinary trends in artificial intelligence, 12th international conference, MIWAI 2018, Hanoi, Vietnam, November 18–20, 2018, Proceedings, vol 11248. Springer, pp 19–31
29. Chaudron D (1995) The "far out" success of teleworking. Superv Manag 40:1–6
30. Bui T, Higa K, Sivakumar V, Yen J (1996) Beyond telecommuting: organizational suitability of different modes of telework. In: Proceedings of the 29th annual Hawaii international conference on systems sciences. IEEE, pp 344–353
31. Hooper D. http://www.cidmcorp.com/wp-content/uploads/2012/10/Telecomuting-and-Workers-Compesnation-CID-Management-White-Paper.pdf. Last accessed 19 March 2018
32. Cross TB. The future technology of working green. http://techtionary.com/books/telecommuting/index.pdf. Last accessed 19 March 2018
33. Gordon GE, Kelly MM (1986) Telecommuting: how to make it work for you and your company. Prentice Hall, Englewood Cliffs, N.J.
34. Bailyn L (1994) Toward the perfect workplace? The experience of home-based systems developers. In: Allen TJ, Scott Morton MS (eds) Information technology and the corporation of the 1990s: research studies. Oxford University Press, New York, NY, pp 410–429
35. Nilles JM (1994) Making telecommuting happen: a guide for telemarketers and telecommuters. Van Nostrand Reinold, New York
36. Di Martino V, Wirth L (1990) Telework: a new way of working and living. Int Labour Rev 129:529–555
37. Doswell A (1992) Home alone?–Teleworking. Management Services, October 18–21
38. Fritz ME, Higa K, Narashiman S (1995) Toward a telework taxonomy and test for suitability: a synthesis of the literature. Group Decis Negot Support 4(4)
39. Gray M, Hodson H, Gordon G (1993) Teleworking explained. Wiley, Chichester
40. Cisco, MobileWorkforce. https://newsroom.cisco.com/dlls/2007/eKits/MobileWorkforce_071807.pdf. Last accessed 19 March 2018
41. ECaTT, Benchmarking progress on new ways of working and new forms of business across Europe (2000). https://web.fhnw.ch/personenseiten/najib.harabi/publications/books/benchmarking-progress-of-telework-and-electronic-commerce-in-europe. Last accessed 19 March 2018
42. Byer J. Apple and Cisco bring mobile workers closer together. https://www.softchoice.com/blogs/advisor/client/apple-and-cisco-bring-mobile-workers-closer-together. Last accessed 14 February 2019

43. Baarne R, Houtkamp P, Knotter M (2010) Het nieuwe werken ontrafeld [Unraveling new ways of working]. Koninklijke Van Gorcum/Stichting Management Studies, Assen, The Netherlands
44. Ten Brummelhuis LL, Bakker AB, Hetland J, Keulemans L (2012) Do new ways of working foster work engagement? Psicothema 24:113–120
45. Grint K (2005) The sociology of work. Polity Press, Cambridge
46. Pettinger L (2005) Friends, relations and colleagues: the blurred boundaries of the workplace. In: Parry J, Taylor R, Pettinger L, Glucksman M (eds) A new sociology of work. Blackwell Publishing, London
47. DeSanctis G (1984) Attitudes towards telecommuting: Implications for work at home programs. Inf Manag 7(3):133–139
48. Baruch Y, Nicholson N (1997) Home, sweet work: requirements for effective home working. J Gen Manag 23(2):15–30
49. Timbal A, Mustabsat A (2016) Flexibility or ethical dilemma: an overview of the work from home policies in modern organizations around the world. Human Resour Manag Int Dig 24(7)
50. Reinsch NL (1997) Relationship between telecommuting workers and their managers: an exploratory study. J Bus Commun 34(4):343–369
51. Thomson P (2008) The business benefits of flexible working. Strateg HR Rev 7(2):17–22
52. Illegems V, Verbeke A (2004) Telework, what does it mean for management? Long Range Plann 37(4):319–334
53. Halford S (2005) Hybrid workspace: re-spatialisations of work, organisation and management. N Technol Work Employ 20(1):19–46
54. Tremblay DG (2003/2004) Balancing work and family with telework? Organizational issues and challenges for women and managers. Women Manag Rev (MCB Press, Manchester) 17:155–170
55. Pool I (1990) Technologies without boundaries: on telecommunications in a global age. Harvard University Press, Cambridge, MA
56. Sidle S (2008) Do people at work have the reputation they deserve? Acad Manag Perspect 22(3):109–110
57. Hamilton CA (1981) Telecommuting. Pers J 66:90–101
58. Duxbury L, Neufeld D (1999) An empirical evaluation of the impacts of telecommuting on intra-organizational communication. J Eng Technol Manag 16:1–28
59. EPI.SK, Z. c. 311/2001 Z.z., 2 cast, pracovny pomer. http://www.epi.sk/zz/2001-311#cast2. Last accessed 20 March 2018
60. Institutzamestnanosti, Telepraca. https://www.iz.sk/sk/projekty/telework. Last accessed 20 March 2018
61. Kordosova M. Telework in Slovakia. https://www.eurofound.europa.eu/observatories/eurwork/articles/telework-in-slovakia. Last accessed 19 March 2018
62. Statista, Percentage of employees that are able to do telework or ICT mobile work in the European Union in 2015, by country. https://www.statista.com/statistics/879251/employees-teleworking-in-the-eu/. Last accessed 14 February 2019
63. ORF.at, Bei Heimarbeit auf Platz vier in EU. https://oesterreich.orf.at/stories/2919988. Last accessed 14 February 2019
64. OECD, Temporary employment. https://data.oecd.org/emp/temporary-employment.htm#indicator-chart. Last accessed 14 February 2019
65. Internetlivestats, Slovakia Internet Users (2016). http://www.internetlivestats.com/internet-users/slovakia/. Last accessed 14 February 2019
66. Kugelmass J (1995) Telecommuting: a manager's guide to flexible work arrangements. Jossey-Bass Inc., US
67. Markarian M (2007) Remote access. Career World 36(2):19–21

68. Noble DF. Digital diploma mills: the automation of higher education. https://journals.uic.edu/ojs/index.php/fm/article/viewArticle/569/490. Last accessed 14 February 2019
69. Beno M (2021) The advantages and disadvantages of E-working: an examination using an ALDINE analysis. Emerg Sci J 5:11–20
70. Beno M (2021) Analysis of three potential savings in E-Working expenditure. Front Sociol 6:675530

Chapter 4
Data and Systems Heterogeneity: Analysis on Data, Processing, Workload, and Infrastructure

Roxana-Gabriela Stan, Catalin Negru, Lidia Bajenaru, and Florin Pop

4.1   Introduction

This paper is our survey toward a general understanding of the requirements for handling large volumes of heterogeneous data, and moreover, presents an overview of the employed computing techniques and technologies necessary for analyzing and processing those datasets. As of our attempt to picture how the data heterogeneity meets the systems heterogeneity, we summarize the identified key issues for multiple dimensions, including data, processing, workload, and infrastructure.

4.2   Data Types, Formats, and Models

In this section, we focus exclusively on data and we enumerate the possible data types to work with, the existing data formats together with different data models.

Given the used techniques for data gathering, data can be classified as follows:

– *observational*—in situ data, gathered by inspecting a subject or an action, observed by humans or collected from sensors.

R.-G. Stan · C. Negru (✉) · F. Pop
University Politehnica of Bucharest, Bucharest, Romania
e-mail: catalin.negru@cs.pub.ro

R.-G. Stan
e-mail: roxanagabrielastan@gmail.com

F. Pop
e-mail: florin.pop@cs.pub.ro; florin.pop@ici.ro

L. Bajenaru · F. Pop
National Institute for Research and Development in Informatics (ICI), Bucharest, Romania
e-mail: lidia.bajenaru@ici.ro

© Springer Nature Switzerland AG 2021
F. Pop and G. Neagu (eds.), *Big Data Platforms and Applications*,
Computer Communications and Networks,
https://doi.org/10.1007/978-3-030-38836-2_4

It is almost impossible to regenerate this real-time data if lost.
- *experimental*—resulted from conducted research studies and performed measurements.
Data can be regenerated if lost, but this is a costly operation.
- *simulated* or *computational*—produced by computerized models designed to emulate how theoretical and real systems would perform given specific parameters.
Data can be safely recovered using the available created models.
- *derived*—data resulted after several transformations applied to the existing data usually originating from multiple sources.
Data recovery is possible, but this is a time-demanding operation.

The varying data types, heterogeneous data sources and formats cannot be handled by the traditional mechanisms. Dedicated data formats and models are continuously introduced in return to the emergent Big Data requirements for computing and storage. When framed through certain use cases, such formats show their strengths and weaknesses. Even though specific algorithms, data mining methods, for example, expect an exact format for the input data, we observe a rising trend in developing different data formats for each new or existing framework.

Considering its representation forms, data can be split into the following categories:

- structured

  • human-generated: recorded clicks, survey entered input, gaming data
  • machine-generated: data from sensors and medical devices, financial systems, web server logs.

- unstructured

  • human-generated: data from social media networks, website content, emails, text messages, mobile data, text and media files
  • machine-generated: satellite imagery and aerial photography, seismic imagery, atmospheric data, scientific data, traffic and radar data, surveillance multimedia files.

- semi-structured (JSON and XML files)

  • human-generated: presentation files, emails compliant to an internal structure or template, digital photos
  • machine-generated: binary executable files, TCP/IP packets, archive files.

Whilst Big Data frameworks such as Apache HBase [14], Apache Hive [15] use known data models for handling structured data, unstructured data found in various formats poses an open challenge as no predefined data models can be established. Furthermore, storing unstructured data implies higher costs compared to structured data.

We should distinguish between raw and processed data in order to assess the existing data formats. Raw data is represented using standard formats, whereas processed data has columnar formats assigned.

Standard formats consist of

– Plain text and log files holding unstructured data
   For voluminous data stored as raw text, even simple computations such as type conversions introduce a significant overhead.
– JSON, XML, CSV, TSV formats describing semi-structured and structured data
   We pinpoint multiple limitations, including the CSV and TSV formats' lack of support for complex data structures, the absence of column types. Since JSON and XML files can't be split, no parallel processing can be performed on them. A potential workaround to achieve improved efficiency would be to convert the incoming data to Big Data-specific formats.
– Big Data particular formats:

   • SequenceFile is a native Apache Hadoop format, designed to be widely used for MapReduce tasks' input, output, and intermediary data, being available in three distinct formats: with no compression or with block-level or record-level compression. The major drawback is that SequenceFile is language-dependent and for integration purposes only the Java API is available.
   • Avro [2] uses JSON for its schema definition and is language-independent. It is the preferred format for write-intensive workloads since new rows can be swiftly appended.

Regarding the columnar formats implying as expected a column-wise storage but per batch of rows, we indicate two Big Data formats, namely, Parquet [23] and Optimized Row Columnar (ORC) [22]. These are reliable options for read-intensive workflows. ORC was completely designed to give Apache Hive a computing performance boost, so it is not an all-purpose format.

In Table 4.1, we compare the most used data formats by highlighting the main similarities and differences between them. As we observe, selecting the suitable format implies specific trade-offs.

We analyze further several data models, differentiating between schema on write traditional database management systems and schema on read Big Data management systems.

The relational database model exposes features such as unique record entries, normalized data, Structured Query Language (SQL) as the language for formulating queries to manage and store data. Marking down the model's issues, complex queries tend to be overly time-demanding, preliminary steps should be taken before having the de facto data. Broadly speaking, structured data resides in relational databases, but this model is too strict for fitting semi-structured or unstructured data into.

The NoSQL database model, adequate for extremely large heterogeneous datasets, addresses the limitations of the prior model, engaging to meet the low latency and scalability urgent requirements. However, the extensive usage of the NoSQL solutions could lead to increased costs.

Modern data models include

– key-value (the simplest non-relational model)
– document (JSON or XML encoded for semi-structured data access)

Table 4.1 Data formats comparison

Data format	Data	Readability	Splitability and compression	Structure	Schema evolution	Big Data platforms
Text file	Raw	Human readable	No	Row-based (CSV)/key-value pairs (JSON)	No support	Presto
Sequence file	Raw	Machine readable binary	Yes	Key-value pairs	Limited support	Apache Hadoop Apache Spark Presto
Avro	Raw	Machine readable binary	Yes	Row-based	Full support	Apache Kafka Apache Hadoop Apache Druid Apache Spark Apache Hive; Apache Pig Presto
Parquet	Processed	Machine readable binary	Yes	Column-based	Limited support	Apache Impala Apache Hive Apache Drill Apache Spark Apache Pig Presto
ORC	Processed	Machine readable binary	Yes	Column-based	Limited support	Apache Hive Apache Pig Apache Spark Presto

Table 4.2 Database systems' supported data models

Engine	Relational	Column	Key-value	Document	Graph	Multi-model
BigTable	–	*	–	–	–	–
Apache Cassandra	–	*	–	–	–	–
Apache HBase	–	*	–	–	–	–
Apache Ignite	*	–	*	–	–	*
Amazon DynamoDB	–	–	*	*	–	*
Mongo	–	–	–	*	–	–
Apache Giraph	–	–	–	–	*	–
ArangoDB	–	–	*	*	*	*
OrientDB	–	–	*	*	*	*

- graph (schemaless, performing poorly for queries that require complete graph traversal)
- object
- tabular
- tuple
- sparse array
- compressed sparse row or column [1].

Multi-model repositories approach data management in a cumulative way by aggregating both relational and document, key-value, graph NoSQL data models [24]. In an effort to manage varied heterogeneous data, several database systems started adopting this multi-model paradigm. Concrete examples are OrientDB and ArangoDB [18].

Table 4.2 exemplifies the most popular NoSQL database engines, highlighting for each of them the supported data models, revealing therefore their multi-model capabilities if applicable.

4.3 Processing Models and Platforms

This section focuses on the computational models and various Big Data systems considering the decentralized processing evolving trends.

The key models employed for computing and analyzing entry data are batch and real-time stream processing. We provide a summary description of these paradigms, as follows:

– *Batch processing*, a computing model that basically enables the conversion of extremely large, finite quantities of static data into batches of jobs in order to be processed in a parallel and distributed manner, returning the results at a later time, once the processing of the entire data collection is done. These batch-oriented computations are automatically correlated with the pioneering MapReduce model [10], with guaranteed scalability.
Rather regarded as a bottleneck, the processing of massive, voluminous data might require extended times making the batch processing inappropriate when the computation times are vital, being, therefore, best suited for workloads with no imposed time constraints for their execution.
– *Real-time processing*, which implies handling received data immediately as well as making available promptly the computed data. An alternate definition for this paradigm refers to the in-memory cluster computing mode enabled especially for delivering swift responses considering the compulsory tight deadlines [5].
– *Stream processing*, which operates on continuous, dynamic data. Contrary to the batch model, stream processing excels at providing low latencies. Data is processed as it flows through the system so quasi-real-time computations are properly ensured. The datasets consist of unbounded streams and only the quantity of the data handled so far is known. Analytics, determining trends over time are adequate applications for the stream-oriented paradigm.
– *Hybrid model*, a unified vision of handling the captured mixed workflows, both stream and batch, aiming to generally address their computing requirements.

These data-centric computing paradigms justify the advent of various distributed processing frameworks. Each platform has trade-offs attached, and for this reason, we highlight in this ongoing study the specific cases when one platform is preferred over another.

In order to operate efficiently, platforms internally make use of distributed data storage systems such as Google File System (GFS) [12] and Hadoop Distributed File System (HDFS).

Based on the data that the computing frameworks are able to process, either in batches or streams or both of them, we classify the platforms into the following categories and group them accordingly:

– batch-only: Apache Hadoop, HTCondor
– stream-only: Apache Storm, Apache Ignite, Apache Heron
– hybrid: Apache Samza, Apache Spark, Apache Flink, Spring XD.

In Table 4.3, we present a comparison of the most frequently used data-intensive platforms and a detailed review follows afterward.

– **Apache Hadoop**
To process batched Big Data, Apache Hadoop uses HDFS (file system), YARN (resource manager, job scheduler), MapReduce (processing engine) as its core components [13]. Key-value data goes through the map, shuffles, reduces execution phases. To enhance or extend beyond the platform's capabilities for data access,

Table 4.3 Comparison of the Apache computing platforms

Platform	Batch mode	Stream mode	Processing guarantee	State management
Apache Hadoop	*	–	Exactly once	stateful
Apache Storm	–	Native, one at a time/micro-batching (Trident)	At least once/exactly once (Trident)	Stateless
Apache Samza	*	Native, one at a time	At least once	Stateful
Apache Spark	*	Micro-batching	Exactly once	Stateful
Apache Flink	*	Native, one at a time	Exactly once	Stateful

management and analysis, several Apache frameworks such as HBase, Hive, Pig could be added to the Hadoop stack or run alongside it.

Hadoop's existing issues include:

- Even though the map-reduce programming model is clearly formulated, developers still need to manually code the data transformations.
- The I/O latencies caused by the local writes and reads of the intermediary computed data, instead of preserving it in memory, directly affect the overall processing speeds.

– **Apache Storm**

Apache Storm is considered, metaphorically speaking, the real-time processing version of Hadoop, remarkable for its facile usage, any programming language support and incredibly fast computations [29]. Storm can grant that the incoming data streams are processed at least one time, and for the Trident abstraction created on top of it, exactly once [31]. Storm keeps running a topology of spouts and bolts, created based on an unlimited stream of processing units called tuples [30].

Storm's unavoidable issues include:

- The order of processing input streams is arbitrary.
- Trident's micro-batching processing shows performance limitations compared to Storm.

– **Apache Samza**

Apache Samza is another stream processing platform, natively designed to integrate with Apache Kafka [25]. Similar to Apache Storm, this platform serves for real-time processing purposes and provides assurance that loss of data in case of failures cannot happen, but duplicates might exist. Streams of data are partitioned for ensured scalability, holding messages as the processing units [26].

Samza's existing limitations include

- Samza is tightly coupled to Apache Kafka.
- Samza yields less flexibility than Storm considering its restricted JVM languages support.

– **Apache Spark**
Apache Spark is incredibly faster compared to Apache Hadoop and could run either on Apache Mesos, Kubernetes, Apache YARN, or standalone. Spark has an impressive collection of built-in libraries, including SparkSQL, GraphX, MLlib [27]. The system serves for both batch and stream computing purposes, enabling the processing of considerably large combined datasets [36]. Spark's in-memory stream processing model uses micro-batches of Resilient Distributed Datasets (RDDs) [35].
Spark's related issues refer to

- Spark has attached higher costs than Hadoop since it is delivered as an in-memory solution, requiring specialized hardware and also demanding more resources.
- As a matter of fact, received data could be processed in quasi-real-time given the micro-batching operations.

– **Apache Flink**
Apache Flink appears as a versatile platform for diverse workloads. The platform supports native, pipelined stream processing. Even for batch processing, Flink treats the static datasets as unbounded message streams [4].
The potential Flink framework's issues refer to

- Flink's late launch in the Big Data processing scenery affected its popularity and widespread adoption.
- It is not regarded as a mature enough platform.

Another in-memory real-time computing platform is Apache Ignite. Evaluated on several applications, Apache Ignite fails to achieve the performance results provided by Apache Spark [28].

By wrapping up the functionalities of the previously discussed Big Data processing technologies and subsequently exposing new interfaces, other useful Apache platforms include

– Flume, Sqoop, Chukwa for data acquisition
– Pig, Hive for data warehouse, data analysis
– Kafka for data streaming
– Mahout, MLlib, SystemML for machine learning.

Furthermore, a valuable addition to this processing heterogeneity analysis is the discussion about opportunistic computing, a paradigm that leverages the benefits of decentralized opportunistic communication between mobile nodes connected wirelessly. Such a dynamic topology is particular for opportunistic networks (ONs), which derive from mobile ad hoc networks (MANETs) [20]. Anyone might wonder what opportunistic computing is all about. It consists of distributed processing using a pool of heterogeneous resources shared by mobile devices. In order to avoid the quick draining of the device's battery, the viable approaches are mobile data offloading and opportunistic dissemination. Replicating or porting data to the mobile nodes found opportunistically the load is reduced. New designs of the underlying algorithms and techniques are of great interest to researchers at the moment [34].

4.4   Workload Types

Based on the stressed subsystems and according to the applications' varying resource demands and consumption, workloads can be characterized as

– CPU-intensive
– memory-intensive
– network or disk I/O-intensive.

The CPU-intensive and CPU-bound terms are used interchangeably across this section. Similarly, the same equivalence applies to the memory and I/O components as well. With regard to the I/O-bound workloads, an even more granular division distinguishes between disk write and read, respectively, network transmit and receive behaviors [3].

The processing and the effective exploit of co-existing heterogeneous workloads is a major challenge. To achieve the fairly desirable performance, the computation, memory, network, and disk-related bottlenecks should be alleviated. This chapter reviews and evaluates various types of data-intensive computing workloads.

Firstly, we enumerate several use cases of real-world workloads, including speech recognition, image processing, face detection, scientific computing, seismic processing, multimedia content streaming, involving the usage of machine learning, search engines, social networks, recommendation systems.

In Table 4.4, we included the studied benchmarked workloads, grouped by their type and decomposed according to the underlying CPU, memory, and I/O-bound tasks.

Table 4.4 Benchmarked workloads

Workload type	CPU-intensive	Memory-intensive	I/O-intensive
Big Data analytics	Sort	K-means	Sort
	Terasort		TeraSort
	PageRank		Grep
	Naive Bayes		PageRank
	K-means		WordCount
			Naive Bayes
High performance computing	HPL Linpack	Sysbench	b_eff_io
	FFTW		bonnie++

Large-scale workloads include the following:

- Big Data analytics workloads
Big Data analytics workloads represent a worth reviewing subset of the modern data center workloads. Since the volume of data loaded by each component of the system differs considerably, it is necessary to analyze certain workloads.
Iconic data analytics workloads include Naive Bayes, SVM, PageRank, Sort, Tera-Sort, Grep, WordCount, K-means, Fuzzy K-means, each of them requiring a distinct resource utilization [16]. To classify the benchmarks as seen in Table 4.4, the workloads were processed on several platforms, such as Apache Hadoop and Apache Spark. We agree that despite their diversity, such workflows are mainly CPU-intensive and I/O-intensive, and could be memory-intensive [19]. The key to performing data analytics often resides in applying adequate data mining and machine learning transformations.
- Machine learning workloads
While there is a large plethora of machine learning algorithms and corresponding applications, we found it very enlightening to outline several case studies which expose diverse workloads that we are interested in.
The research work from [21] targets high-dimensional, sparse datasets containing unstructured text and introduces a template for accelerating this type of machine learning workloads.
The training jobs of deep learning workloads are either memory-intensive or compute-intensive as described in [32]. An important consideration is that GPUs executing deep neural networks can bring an additional boost of performance [17]. Based on these recent advances, we can consider that the machine learning workloads are commonly CPU-intensive and memory-intensive, occasionally I/O-intensive.
Even though some of the aforementioned platforms offer distributed machine learning algorithms support, if employed on Apache Spark, for instance, considerable performance drops are reported [11].
- Sparse matrix workloads
When it comes to large matrices, either dense or sparse, obviously a significant amount of memory is needed. These matrices could not always fit into memory. The vast majority of the large matrices used in practice are sparse, namely, they are filled almost entirely with zeroes and contain just a few actual data values.
Basic linear algebra operations such as multiplication of matrices with high dimensionality and sparsity characteristics, demand to compute resources of CPU and GPU heterogeneous platforms, accessed in a distributed manner. The split of such a matrix generates hypersparse submatrices that need to be handled accordingly.
Working with raw sparse matrices leads to memory and computation resources wasted on handling mostly zero values. Of the formats previously discussed, compressed sparse row or column could be more efficient for representing data in this context. The sparse matrix workloads are therefore memory-intensive, CPU-intensive and I/O-intensive.

Use cases include machine learning as many of the learning methods rely on sparse array computations, recommender systems, social networks for maintaining a friendship graph using sparse matrix operations, computer vision techniques for extracting features from pictures mostly made of black pixels.
- Numerical methods workloads in High-Performance Computing
  Numerical methods including Monte Carlo and the finite element method can be applied to solve important equations, in particular evolution equations such as Navier-Stokes. These equations can be used in computational fluid dynamics, for building climate models, or for achieving a better weather forecast accuracy. Accordingly, the numerical methods workloads are CPU-intensive, I/O-intensive, and memory-intensive.

4.5   Infrastructure Types

This section addresses the infrastructure heterogeneity by presenting a brief overview of the key network-based computing paradigms.

With the Internet of Things (IoT) continuous development and growth, and subsequently the massive repositories of generated data, there is an emergent need to reduce the gap between mobile and cloud computing. Depending on where and how the processing happens at infrastructure specific levels, we differentiate between cloud, fog, edge, mobile cloud, and drop computing.

Table 4.5 highlights the main identified benefits and drawbacks of each infrastructure type that we present in this regard.

- **Cloud computing**
  The cloud computing paradigm makes use of data centers for managing data and adopts a pay-as-you-go pricing model, applied by various providers including Google, Amazon, or Microsoft. It is the most widely used paradigm nowadays, leveraging all the types of clouds, namely public, private, or hybrid, in all flavors (IaaS, PaaS, SaaS).
- **Fog computing**
  Compared to cloud computing, this paradigm employs networking, computation and storage in the vicinity of the IoT devices, being introduced to address the issues of the former one [33].
- **Edge computing**
  The edge computing paradigm migrates the computation at the location where the data is generated by sensors and IoT devices, being especially suitable for real-time applications.
- **Mobile cloud computing**
  This paradigm combines the complementary capabilities of mobile and cloud computing, as already suggested by the name, moving most of the computations outside of the mobile device in an effort to save its battery [33]. Crowdsourcing

Table 4.5 Infrastructure comparison

Infrastructure	Advantages	Disadvantages
Cloud computing	– High availability – High processing capacity	– Doesn't scale with the costs – High latencies on distant clouds – Bandwidth limitations – Security and privacy breaches – Poor customizations for public clouds – High power consumption – High costs to build the infrastructure – Internet connection is required
Fog computing	– Flexibility – Guarantees mandatory low latencies, high bandwidth, privacy and security – Decentralization – Power consumption lower than cloud – Internet connection is not always required – Relatively close to users – Real-time application support – Mobility support	– Moderate availability – Limited scalability
Edge computing	– Wide range of connectivity types – Low power consumption – Latencies usually lower than cloud and mobile cloud – Internet connection is not always required – Close to users – Real-time application support – Mobility support	– Limited processing capacity – Limited storage capacity
Mobile cloud computing	– High availability – High processing capacity	– High latencies – Internet connection is required – High device battery consumption – Unsustainable model in the future
Drop computing	– Decentralization – Internet connection is not always required – Mobility support – Available for a huge number of mobile devices – Efficient data handling (data offloading) – Opportunistic mobile communication – Reduced risks of security and privacy breaches – Costs much lower than mobile cloud – Enables horizontal scaling	– Consistency issues – Considerable device battery consumption

applications in various domains such as health care, environmental monitoring are feasible use cases.

– **Drop computing**

Drop computing [9] is a novel paradigm that addresses the limitations of mobile cloud computing by proposing a multilayered architectural solution. Basically, it combines opportunistic networks in the first place enabling the ad hoc communication between nearby mobile devices, with edge and fog nodes and further cloud in order to offload data and the complex compute-intensive tasks [6, 8]. As this research work is at its early stages, the potential issues such as data corruption are gradually addressed [7].

4.6 Conclusion

This paper presented a four-dimensional analysis to explain data and systems heterogeneity. We have highlighted the existing limitations for the computing platforms, data models, workloads, and infrastructures that we have considered in our study. We conclude that it is desirable to allow the scalable processing of heterogeneous workloads, holding heterogeneous data into an all-purpose platform.

References

1. Ahmed S, Usman Ali M, Ferzund J, Atif Sarwar M, Rehman A, Mehmood A (2017) Modern data formats for big bioinformatics data analytics. Int J Adv Comput Sci Appl 8(4):366–377
2. Avro, Apache Software Foundation. https://avro.apache.org/
3. Azmandian F, Moffie M, Dy JG, Aslam JA, Kaeli DR (2011) Workload characterization at the virtualization layer. In: 2011 IEEE 19th annual international symposium on modelling, analysis, and simulation of computer and telecommunication systems, pp 63–72
4. Carbone P, Katsifodimos A, Ewen S, Markl V, Haridi S, Tzoumas K (2015) Apache Flink™: stream and batch processing in a single engine. IEEE Data Eng Bull 38(4):28–38
5. Casado R, Younas M (2015) Emerging trends and technologies in big data processing. Concurr Comput Pract Exp 27(8):2078–2091
6. Ciobanu R, Dobre C, Bălănescu M, Suciu G (2019) Data and task offloading in collaborative mobile fog-based networks. IEEE Access 7:104405–104422
7. Ciobanu R, Tăbuşcă V, Dobre C, Băjenaru L, Mavromoustakis CX, Mastorakis G (2019) Avoiding data corruption in drop computing mobile networks. IEEE Access 7:31170–31185
8. Ciobanu R-I, Dobre C (2019) Mobile interactions and computation offloading in drop computing. In: Advances in network-based information systems. Springer International Publishing, pp 361–373
9. Ciobanu R-I, Negru C, Pop F, Dobre C, Mavromoustakis CX, Mastorakis G (2019) Drop computing: ad-hoc dynamic collaborative computing. Futur Gener Comput Syst 92:889–899
10. Dean J, Ghemawat S (2004) MapReduce: simplified data processing on large clusters. In: Sixth symposium on operating system design and implementation, OSDI'04, San Francisco, CA, pp 137–150
11. Dünner C, Parnell T, Atasu K, Sifalakis M, Pozidis H (2017) Understanding and optimizing the performance of distributed machine learning applications on apache spark. In: 2017 IEEE international conference on big data (big data), pp 331–338

12. Ghemawat S, Gobioff H, Leung S-T (2003) The Google file system. ACM SIGOPS Oper Syst Rev 37(5):29–43
13. Hadoop, Apache Software Foundation. https://hadoop.apache.org/
14. HBase, Apache Software Foundation. https://hbase.apache.org/
15. Hive, Apache Software Foundation. https://hive.apache.org/
16. Jia Z, Zhan J, Wang L, Luo C, Gao W, Jin Y, Han R, Zhang L (2017) Understanding big data analytics workloads on modern processors. IEEE Trans Parallel Distrib Syst 28(6):1797–1810
17. Lew J, Shah DA, Pati S, Cattell S, Zhang M, Sandhupatla A, Ng C, Goli N, Sinclair MD, Rogers TG, Aamodt TM (2019) Analyzing machine learning workloads using a detailed GPU simulator. In: 2019 IEEE international symposium on performance analysis of systems and software (ISPASS), pp 151–152
18. Lu J, Irena H (2019) Multi-model databases: a new journey to handle the variety of data. ACM Comput Surv 52(3)
19. Mohammadi Makrani H, Sayadi H, Pudukotai Dinakarra SM, Rafatirad S, Homayoun H (2018) A comprehensive memory analysis of data intensive workloads on server class architecture. In: Proceedings of the international symposium on memory systems, MEMSYS'18, New York, NY, USA. Association for Computing Machinery, pp 19–30
20. Marin R-C, Ciobanu R-I, Dobre C (2017) Improving opportunistic networks by leveraging device-to-device communication. IEEE Commun Mag 55(11):86–91
21. Mishra AK, Nurvitadhi E, Venkatesh G, Pearce J, Marr D (2017) Fine-grained accelerators for sparse machine learning workloads. In: 2017 22nd Asia and South Pacific design automation conference (ASP-DAC), pp 635–640
22. ORC, Apache Software Foundation. https://orc.apache.org/
23. Parquet, Apache Software Foundation. https://parquet.apache.org/
24. Płuciennik E, Zgorzałek K (2017) The multi-model databases: a review. In: Beyond databases, architectures and structures. Towards efficient solutions for data analysis and knowledge representation. Springer International Publishing, pp 141–152
25. Samza, Apache Software Foundation. https://samza.apache.org/
26. Samza–Core concepts, Apache Software Foundation. http://samza.apache.org/learn/documentation/latest/core-concepts/core-concepts.html
27. Spark, Apache Software Foundation. https://spark.apache.org/
28. Stan C-S, Pandelica A-E, Zamfir A-V, Stan R-G, Negru C (2019) Apache spark and apache ignite performance analysis. In: 2019 22nd international conference on control systems and computer science (CSCS), pp 726–733
29. Storm, Apache Software Foundation. https://storm.apache.org/
30. Storm–Concepts, Apache Software Foundation. https://storm.apache.org/releases/current/Concepts.html
31. Storm–Guaranteeing Message Processing, Apache Software Foundation. https://storm.apache.org/releases/current/Guaranteeing-message-processing.html
32. Wang M, Meng C, Long G, Wu C, Yang J, Lin W, Jia Y (2019) Characterizing deep learning training workloads on Alibaba-PAI. In: 2019 IEEE international symposium on workload characterization (IISWC), pp 189–202
33. Yousefpour A, Fung C, Nguyen T, Kadiyala K, Jalali F, Niakanlahiji A, Kong J, Jue JP (2019) All one needs to know about fog computing and related edge computing paradigms: a complete survey. J Syst Architect 98:289–330
34. Yu S, Zhang L, Li L, Yan B, Cai Z, Zhang L (2019) An efficient interest-aware data dissemination approach in opportunistic networks. Procedia Comput Sci 147:394–399
35. Zaharia M, Chowdhury M, Das T, Dave A, Ma J, McCauley M, Franklin MJ, Shenker S, Stoica I (2012) Resilient distributed datasets: a fault-tolerant abstraction for in-memory cluster computing. In: Proceedings of the 9th USENIX conference on networked systems design and implementation, NSDI'12, USA. USENIX Association
36. Zaharia M, Xin RS, Wendell P, Das T, Armbrust M, Dave A, Meng X, Rosen J, Venkataraman S, Franklin MJ, Ghodsi A, Gonzalez J, Shenker S, Stoica I (2016) Apache Spark: a unified engine for big data processing. Commun ACM 59(11):56–65

Chapter 5
exhiSTORY: Smart Self-organizing Exhibits

Costas Vassilakis, Vassilis Poulopoulos, Angeliki Antoniou, Manolis Wallace, George Lepouras, and Martin Lopez Nores

Abstract Creating stories for exhibitions is a fascinating and in parallel laborious task. As every exhibition is designed to tell a story, museum curators are responsible for identifying, for each exhibit, its aspects that fit to the message of the story and position the exhibit at the right place in the story thread. In this context, we analyze how the technological advances in the fields of sensors and Internet of Things can be utilized in order to construct a "smart space," which consists of self-organizing exhibits that cooperate with each other and provide visitors with comprehensible, rich, diverse, personalized, and highly stimulating experiences. We present the system named "exhiSTORY" that intends to provide the appropriate infrastructure to be used in museums and places where exhibitions are held in order to support smart exhibits. The architecture of the system and its application potential is presented and discussed.

C. Vassilakis · A. Antoniou · G. Lepouras
Department of Informatics and Telecommunications, University of Peloponnese,
Tripoli, Greece
e-mail: costas@uop.gr

A. Antoniou
e-mail: angelant@uop.gr

G. Lepouras
e-mail: gl@uop.gr

V. Poulopoulos · M. Wallace (✉)
Knowledge and Uncertainty Research Laboratory, University of Peloponnese, Tripoli, Greece
e-mail: wallace@uop.gr

V. Poulopoulos
e-mail: vacilos@uop.gr

M. L. Nores
University of Vigo, Vigo, Spain
e-mail: mlnores@det.uvigo.es

© Springer Nature Switzerland AG 2021
F. Pop and G. Neagu (eds.), *Big Data Platforms and Applications*,
Computer Communications and Networks,
https://doi.org/10.1007/978-3-030-38836-2_5

5.1 Introduction

An exhibition, which is usually formed by a number of exhibits, is not a simple set of objects randomly selected and placed altogether. In reality, an exhibition is designed to "narrate a story" [35]. The procedure of selecting exhibits and designing stories to be told by them is a tedious task that requires highly trained curators [24] as they are responsible for a large part of a museum's budget [25]. These issues lead museums to utilize only a portion of the exhibits that they own [22], while the presented number of stories that the selected exhibits can narrate may also be limited.

Theoretically, the work of the curator is to organize exhibits within a limited amount of space in order to describe a story [20]. Practically, the analysis of objects and the extraction of a common story is a tedious procedure, as the curator needs to identify the links between the exhibits as well as the description of a definite and meaningful story. On the other hand, the visitors of a place need to be provided with information that reveal the "hidden" story of the exhibition, which is typically conveyed using multimedia information, e.g., text, audio, video, and images. However, even with application of all the aforementioned procedures, this way of exhibition setup is able to provide only a limited number of views to the visitors, and more specifically can stimulate only the visitors that are able to recognize at least one of the stories woven by the curator—and are also interested in at least one of them.

As a matter of fact, every single object of an exhibition has many different stories to tell, for example, Titian's "Diana and Callisto" (Fig. 5.1) [12] can tell us about ancient Greek gods; about deception wrath and humiliation; nudity in art; about the artist's personal style or the trends of the artistic period; and so forth. It is obvious that from a single piece of art it is possible to extract a number of different views; when more exhibits are combined then it is expected that the combination of stories is countless, for instance, when the painting of Diana and Callisto is next to The Arnolfini Portrait (Fig. 5.2 [36]) a new set of stories and connections can emerge, like wanted and unwanted pregnancies (both show pregnant women in different situations), maternity practices, etc. Similarly, if all three are connected, then new connections can be revealed like women's rights, European art, social practices around female appearance, etc. Despite that, only few stories are selected to be presented in every exhibition.

This is the reason a number of projects such as PEACH [34], HyperAudio, and HIPS [30] combine the content of the museum and the context under which an exhibit is viewed in order to construct context-aware museum guides. In this manner, they are able to tell automatically synthesized stories utilizing narration generation algorithms. The generated stories can be further refined or filtered to match the user interests, the visitor model, the interaction history, or any other context parameter. These works are based on the fact that exhibits are static objects placed "forever" in a place combined with a set of similar objects. As such, when new exhibits are introduced, or if an object is removed or placed in other location, the whole procedure of narration generation must be reproduced in order to match the new setup. That being told, there is a strong need for manual work to be done which in most of the

Fig. 5.1 Titian's "Diana and Calisto"

Fig. 5.2 Jan Van Eyck's "The Arnolfini Portrait"

cases is laborious. Furthermore, the previous works imply that the curators should be able to record the metadata of each of the objects to be included to the automatic narration generation procedure; this means that they must have knowledge of deep semantic content, which in most of the case is trivial and usually does not include unexpected relationships between exhibits.

Going a step back we examine the research area of Internet of Things (IoT) which is flourishing the last years. The landscape of IoT is getting thoroughly analyzed, meaning that communication technologies, discovery functionalities, and mechanisms are studied in detail [6].

In [8, 9], the idea of the single smart space concept and its capabilities is defined. They actually represent a new way of realizing the concept of smart spaces within the cultural heritage domain. Similarly to what we present in our work, a form of "smart exhibit" is presented. A *server* smart exhibit may retrieve multimedia files from *client* exhibits, but apart from data exchange there is no description of self-organized exhibition or personalized narration based on the architecture of the IoT-based system. Indoor cultural activities have been also studied within the framework of the IoT and significant factors that affect the museum visits have been identified, like the visit context [17]. Another example of museum IoT system is the iPhone App, "Take me I'm yours." The project explores ways that objects can talk to the visitors and require actions [33].

Discussing about mobility, as visitors usually are "moving objects" into a smart space, we need to rely on this aspect when analyzing cultural heritage visits [41]. IoT research is coupled with mobile devices as a way to assist people in their everyday lives. Current research is focusing on unique ways to interact with appliances in the surrounding environment with the user's mobile devices [11]. One of the most known and award winning works regarding IoT, mobility, and cultural heritage was the QRator project applied at the UCL Grant museum of Zoology. Internet-linked interactive museum item labels were used to construct narratives and increase visitor engagement [21].

In this paper, we present the transformation of plain objects to *smart exhibits*. Smart exhibits are able to have "knowledge" of their own stories, communicate with each other, are able to self-organize as well as offer more than a simple presentation of themselves but also provide rich, diverse, and highly stimulating experience to the visitors. In that scope we develop exhiSTORY (from the Greek word $\epsilon\xi\iota\sigma\tau o\rho\epsilon\acute{\iota}$: tells a story), a framework that allows for self-aware exhibits positioned within the same smart space, to cooperate and work together, to produce self-organized exhibitions, each one telling a coherent story. Besides information that accompany exhibits in real, exhiSTORY also considers information regarding each individual user, such as interests, visit context, user device capabilities, etc., thus generating tailor-made museum visiting experiences, adjusted to each one's preferences, interests, and style, increasing the overall quality of experience. In this paper, we focus on a special infrastructure that allows exhibits to interconnect and interact with each other and with the visitors.

The rest of this paper is organized as follows: in Sect. 5.2, we discuss the stories that could be told by self-aware and information-rich exhibits, while in Sect. 5.3 we explore different methods for implementing smart exhibits in the context of IoT and discuss how self-organization of exhibitions can be accomplished in each option. Continuing, in Sect. 5.4 we present the architecture of exhiSTORY, the framework that generates the stories to be told and in Sect. 5.5 we describe the system in operation. Finally, we close our discussion in Sect. 5.6 with our concluding remarks.

5.2   The Stories Told by Exhibits

In a conventional museum, each exhibit is accompanied by an information label, usually presenting certain information such as the title, the name of the artist, and the time of creation, possibly complemented with a brief description.

Of course, via the process of duration, the museum holds a lot more information regarding any item, information that typically include the context of the creation, the context of the artist (who taught him, who inspired him), the context of the content (what is depicted, what the artist meant to convey, what other theories exist regarding its meaning or intentions) as well as the history of the exhibit as an item. Various museum information standards, including the Cataloguing Cultural Objects (CCO) standard [7] and the SPECTRUM standard [10] organize this additional information into concrete structures, and describe best practices for populating these structures.

Additional information cannot be presented in a typical setting as there is limited space for detailed information. However, by shifting the narration viewpoint from the whole of the exhibition space to the exhibit and allowing the exhibit to present itself exploiting its own plentiful information and taking into account the context of its surroundings, richer experiences could be offered to visitors. For example, regarding the presentation of its own information, it is now possible for the exhibit "Diana and Callisto" to present to us all the information that could not be displayed in its small accompanying label. However, the full potential of the exhiSTORY system stems from the fact that having access to the full context and semantic information of the exhibits opens up great opportunities, which could be focused on looking for connections between them. Google[1] has experimented with the notion of x degrees of separation of items in museum collections based on their visual similarity [23], but we find the **semantic** notion of x degrees of separation far more stimulating.

In this notion, we suggest that given any pair of items meaningful links between them can be located with a reasonably small number of steps that go through facts related not only to history or art, but also to popular culture and any available information about the target user's memories and interests; experiments reported in [40] provide evidence that user-specific intermediate entities can be used as elements of the path linking two nodes in a semantic network, and we plan to further explore this issue through experiments specifically targeted to the aforementioned aspects (popular culture, user's memories, and interests).

The aforementioned incline to an innovative assumption. The main idea is that every object that is considered to be part of our heritage has knowledge about itself and a character; the knowledge and character is not confined to the standard information that accompany an object (this typically includes the creator—artist—the creation date, the material, and the usage), but also encompasses a whole new set of data that could include information like: what do I mean, with which concepts am I related and under which interpretation, where have I traveled, how many people have seen me, how many generations know me, or what are other objects that we co-existed at the same place. In this manner, large parts of hidden histories will be revealed. In

---

[1] Google Inc. http://google.com.

parallel, the interconnection of each object and the person observing it can be further analyzed, thus allowing experience personalization. Exploitation of the profile can be achieved through social networks, web platforms, or alternate procedures [1, 13, 16, 18].

It is inevitable that advances in technology affected the procedure through which a narration is recognized and presented to visitors. The "PEACH" project uses mobile devices and uses cinematic techniques in order to create a feeling of personalized TV. The documentary-like content also adapted to the interests of the user [32]. Another known system is "HIPS" (Hyper-Interaction within Physical Space), a hypermedia system supporting mobile presentation of museum and historical information. Tourists' positions were detected and auditory information was personalized and context depended [5]. The main principle of the application was that information is context dependent, and thus it should be presented in different ways [30]. The environment became an interface and the visitor's movements became a form of input to the system. HIPS assumed that different visiting styles need different durations for the presentations and the empirical data support this hypothesis [15]. HIPS was using infrared emitters to connect to the users devices (PDAs) [27]. Finally, user testing and evaluation showed that all users liked the idea of receiving information related to their movement. In addition, in the experimental cases where the visiting style was matched to appropriate content, the users demonstrated increased interest by requesting more information about the exhibits explicitly [27].

The *exhiSTORY* system complements the aforementioned approaches to multiple narrative generation and personalization by exploiting IoT technologies, through which exhibits are able to contribute their own semantic information to the venue, coordinate and collaborate to tell a coherent story to a visitor according to her profile. In addition, exhiSTORY provides the relevant infrastructure to materialize this combination.

In this paper, we propose an ad hoc network that includes objects and people. This network will be able to interact uniquely with each person according to parameters that may include among others the enriched content history of an object. The ultimate goal is to calculate the degrees of separation between objects and people, targeting to the maximization of the impact that can exist through the interactions of people and cultural heritage objects; the approach presented in [3] is one possible way to do this.

In the next section, we explore different approaches for implementing smart, self-aware exhibits in the context of IoT and discuss how self-organization of exhibitions can be accomplished in each option.

5.3 The Smart Exhibit

Analyzing the information discussed in Sect. 5.2 we realize that, in order to achieve an automated procedure for creating stories from sets of objects, each object must carry information such as origins, history, meaning, interpretations, and context.

Fig. 5.3 The smart exhibit

Metadata   History   Context

Moreover, there is a strong need for sharing of this information as well as enriching and updating this information either manually or automatically (Fig. 5.3). Different implementation approaches are reviewed below in an attempt to identify the desired characteristics for exhiSTORY's infrastructure. In the following, we review the main options available and examine the strengths and weaknesses of each approach concluding to solutions that can be applied to different types of venues in order to have optimal use.

Elementary implementation. Exhibits carry an RFID tag providing only a single identification number. The RFID tag is sensed by RFID readers hosted in the exhibition rooms, which notify a museum-hosted server regarding the locations of the exhibits [38]. The museum-hosted server maintains a repository containing all the information (descriptions, semantic information, and RFID identifications) for the exhibits that it accommodates, therefore having obtained the exhibit location information from the RFID readers, it can run logic for exhibit presentation and dynamic exhibition formulation. Visitors receive the exhibition and exhibit information by connecting to the server through standard Internet connectivity, typically supported through Wi-Fi access points installed by the museum. Visitor requests to the server may contain visitor profile information, which the server can exploit to perform personalization (Fig. 5.4).

The core benefit of this implementation is its low cost, as only RFID tags and readers are needed. An RFID sensing grid must also be installed in the venue to support automated geolocation of exhibits. More details on the process of automatically geolocating exhibits using RFID tags are provided in Sect. 5.3.2.

On the other hand, the exhibits with this approach do not truly "carry" their information: that information is stored on a museum-hosted server, limiting the scope of the approach to exhibit mobility within a single museum. If an exhibit is moved from one museum to another, the exhibit-related information must be manually entered or imported to the server of the receiving museum.

Security in this implementation option is high, as a malicious visitor can only emulate one of the RFID tags of the exhibition, trying to trick the system to believe that an exhibit is at a place other than its true location (as contrasted to subsequent approaches where malicious parties can try to inject content to the exhibition).

Fig. 5.4 Elementary
implementation

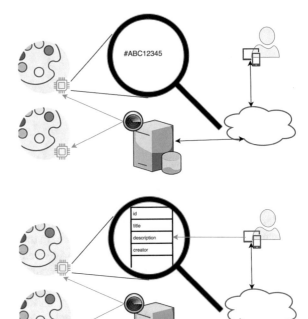

Fig. 5.5 Basic
implementation

A discussion on the security aspects of the different smart exhibit implementations
is provided in Sect. 5.3.3.

Basic implementation. Same as the elementary implementation, but the RFID
chip also has the exhibits' basic information, such as title and creator, as well as
a brief description including key semantic properties. Typically, an RFID tag can
accommodate up to 2 KB of data [31], so the amount of this information is essentially
limited and external storage services are required to store additional data for the
exhibit, including extended semantic data, multimedia files, enhanced descriptions,
histories, and so forth. These storage services will be provided by the museum's
server (Fig. 5.5).

Similar to the elementary implementation, the museum-hosted server undertakes
the exhibit presentation and dynamic exhibition formulation, as well as the delivery
of information to visitors. However, in this option each exhibit "carries" along some
information when it is moved from one museum to another, hence the receiving
museum server can retrieve this information in a plug-and-play fashion and use it
to readily integrate the exhibit to the museum's collections (or even formulate new
collections). Typically though, additional information (such as multimedia files and
extended descriptions) will need to be entered or imported to the museum server
for the exhibit, in order to enable presentation of rich information and enhanced
storytelling features.

Fig. 5.6 Memory-rich
implementation

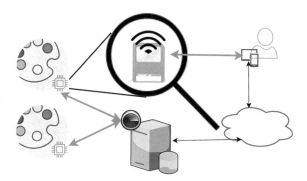

Exhibit geolocation in this case is performed in the same manner as with the elementary implementation and is discussed in Sect. 5.3.2. Regarding security, the risk in this case is higher than the case of elementary implementation, since exhibits carry information and a malicious party can exploit this feature to inject content into the exhibition; a relevant discussion is presented in Sect. 5.3.3.

Both elementary and base implementations are similar to the condition under which is currently the exhibits in the museums with the differentiation of having a mechanism on them has a number of data stored within. No communication can be achieved other than a device scanning the mechanism (RFID in our occasion) in order to obtain this information. This concludes to the fact that the first can even be achieved with some form of ID presentation such as a barcode or a 3D code (e.g., QR code).

Memory-rich implementation. By enhancing exhibits with memory-rich capabilities, there is the option of data exchange and communication, which is a major differentiation with respect to the already presented solutions. Supposing that each exhibit can have large storage so as to "carry" its own information, then what is required is a means of communication. This can be implemented by having *FlashAir* cards [39] installed on them, which may provide from 8 to 32 GB of storage space, plus wireless LAN communication capabilities. These imply that the existing information and any derived or computed information can possibly be stored within the exhibit (Fig. 5.6).

The implementation scheme allows mobility and automation during the procedures of exhibition organization in any place, any time. As the exhibits provide with content the system requires a server that is able to discover exhibits aligned together and perform all the processes that lead to exhibition formulation and presentation of personalized information. The museum-hosted server may store additional data on the exhibit's memory, such as the exhibitions it has participated in or information regarding the profile of the visitors that have viewed it, thus the exhibit's level of self-awareness can be progressively elevated. Exhibit geolocation can be fostered by standardized Wi-Fi triangulation or Wi-Fi fingerprinting methods [4, 28]; alternatively RFID tags can be also attached to exhibits to implement geolocation through an RFID reader antenna grid. Exhibit geolocation is further discussed in Sect. 5.3.2.

Fig. 5.7 Agent
implementation

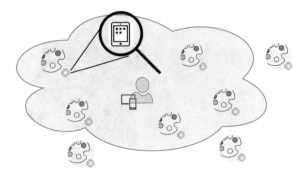

The cost in this case is higher. The main disadvantage to be considered is that exhibits will now require to be connected to a source of power to sustain the operation of the *FlashAir* cards. Security is also an issue to consider, as (1) exhibit storage is writable and (2) visitors are presented with information and media files provided by the exhibits, so a malicious party may try to emulate an exhibit, and thus inject content into the exhibitions. Security risks and mitigation options are discussed in Sect. 5.3.3.

Agent implementation. This implementation includes a small device (e.g., a low-cost embedded device) that accompanies each exhibit. In this occasion, each exhibit acts as an intelligent software agent. Consequently, there is no need for a centralized server as the whole procedure is based on ad hoc networks created by intelligent agents with enough computational power as well as connectivity in order to achieve the system's desired procedures. The levels of independence of each exhibit are such that the system is not limited to indoor settings. Personalization is possible if visitors are able to become peers of the ad hoc exhibition network and share personalization information (Fig. 5.7).

Exhibits can be automatically geolocated, by exploiting their Wi-Fi signal strength, using standardized Wi-Fi triangulation or Wi-Fi fingerprinting methods [4, 28] (c.f. Sect. 5.3.2). Furthermore, GPS units for low-cost embedded devices do not affect much the cost of the overall system. As is often the case with agent-based systems, security risks are highest as a malicious user can join the network of exhibits and participate as a peer in the negotiation and decision-making process, co-shaping the content to be displayed on all exhibits. These security risks can be addressed through the use of public key cryptography, as discussed in Sect. 5.3.3.

5.3.1 Centralized System Control

The implementations presented are intended to provide information about how the environment can be formed and understood; the choice among the implementations is directly related to the exhibits and the exhibitions as well as the capabilities and particularities of each venue. In most of the cases, a local server is responsible for exchanging information with the exhibits. The approach of a local server used to perform all the procedures offers simplicity in terms of technology and algorithms, despite the fact that it may require ample computational power and storage space to accommodate the needs of the exhibits; depending on the storage capabilities of the exhibits and the information stored on them the role of the server is differentiated, as its main role apart of the facilitation of the exchange of information is to store what cannot be stored on the exhibits. Scalability for this system can be provided by exploiting the elasticity feature of standard cloud architectures.

In this work, we will further elaborate on an architecture based on the memory-rich implementation of smart exhibits. The rationale behind choosing the memory-rich implementation of smart exhibits as a demonstrator was based on the following aspects:

- in this implementation exhibits carry with them all their information, making them self-contained objects that are seamlessly integrated into the IoT architecture;
- this implementation uses standard, "off-the-shelf" components (*FlashAir* cards), with an affordable cost, as contrasted with the *agent* implementation, where either custom hardware needs to be built or a small computer, with higher cost than *FlashAir* cards, needs to be used.

According to the aforementioned, we should be able to create a more generic architecture that could support all the implementations described, at least the ones that require less technical setup than the one selected. Of course, we should always pay attention on the capabilities and constraints of the place to be supported by the system as defined in Sect. 5.3.4.

5.3.2 *Automated Exhibit Geolocation*

An important aspect of a sensors system applied on a set of moving objects is based on the definition of the geolocation. Automated exhibit geolocation can be performed by exploiting the RFID tags attached to the exhibits and/or the Wi-Fi adapters which exhibits carry, depending on the smart exhibit implementation adopted by the venue. In the case of the museums, we need a very precise geolocation of each object in order to achieve the best narrative for the visitors. According to [38], RFID tags can be used to geolocate objects, exploiting the signal strength captured by appropriately placed RFID reader antennas. This work asserts that the detection of passive RFID tags provides excellent precision when the distance between the reader antenna and the tag is up to 3 m, and exceeds 75 % for distances up to 4.5 m, hence setting up a 2 m × 2 m rectangular grid of reader antennas will provide adequate precision for geolocating the exhibits. The number of required reader antennas can be, however, significantly reduced by exploiting the fact that when venue exhibitions are restructured, a number of exhibits within the exhibitions are not moved. Therefore, exhibits that remain still can be used as reference tags, and by applying a weighted-center of gravity technique, one reader antenna can be proved sufficient for an area of 12 m by 10 m, providing a level of accuracy of about 1.07 m [19]. This setup is adequate for the positioning the exhibits within the generated narratives (the visitor can be directed to elements of the narrative at exhibition room level or exhibition room area level). These approaches apply to the elementary and the basic implementation for smart exhibits. Note that instead of polling for non-moved exhibits to identify reference tags, extra RFID tags can be positioned in the venue to play this role, i.e., provide reference points for geolocating exhibits.

When smart exhibits are implemented using either the memory-rich or the agent approach, the Wi-Fi adapters carried by the exhibits can be exploited to provide exhibit geolocation. In this case, the museum needs to have available a set of Wi-Fi access points, and then techniques such as Wi-Fi fingerprinting or trilateration [28] can be used to perform exhibit geolocation. Wi-Fi geolocation methods have been successfully used for identifying visitor locations in the Experimedia Blue project [26, 29], while the Art Institute of Chicago (AIC) has also used Wi-Fi-based geolocation to compute the location of visitors reliably within a range of 10 m [37]. If exhibits also carry an RFID tag, RFID readers can be used as an additional source of geolocation information as described above. In case of more than one sources of geolocation information, the more precise is selected to be used.

5.3.3 Security Aspects

Security is a base aspect of every system that includes communication and connection to end users. The complexity of the system negatively correlates with its security, meaning that the simpler the system, the higher the security. In case of elementary and basic implementations, the exhibits either carry only their identification, or their identification along with some limited information about them. As such, the only security issue that can occur is the user emulating the RFID tag, thus tricking the system in believing that the exhibit is placed at a different location than the real one. However, the problem can be detected by examining the uniqueness of each RFID transmitted. As mentioned in Sect. 5.3 in the cases of system setup with dynamic two-way information exchange, there is a possibility of malicious users trying to emulate an exhibit. What one can achieve by emulating an exhibit could be from simply inserting content into the system procedure, to trying to acquire user profiling information. In the case of the *FlashAir* cards, a first precaution measure is to utilize the security features existing within the hardware and the accompanying software in order to prevent emulation or data acquiring or alteration. The basic feature that can be utilized is the configuration of the devices to Wireless LAN client mode [14] and force their setup for network connection to point to a specific device. In parallel, the connection point (Access point) that objects are connected to has to be configured to allow connection only to devices with specific characteristics which will be connected as exhibit content providers. In this way, we are assured that no malicious device can be connected to this network and exchange any kind of information but on the other hand every exhibit added to the collection needs to be carefully setup by a technical expert.

5.3.4 Selecting an Implementation Option for Smart Exhibits

Above, four different implementation options for smart exhibits have been described. In this subsection, we provide some guidelines regarding the selection among the different implementation options. *The cost factor*: Cost is probably the most important factor to consider, since if some implementation option is beyond the budget capabilities of the venue, then it is clearly inapplicable and cannot be considered further. Regarding the cost criterion, the two RFID-based techniques (elementary and basic implementation) have an edge, since the cost of the RFID tags is minimal. Building a dense array of RFID reader antennas can prove costly, especially for large spaces, however exploiting stationary exhibits or reference RFID tags to assist in geolocating other exhibits can significantly reduce the cost. The memory-rich implementation is ranked next, with a final cost of the complete setup described ranging from $35 to 37. Finally, the agent-based implementation is the most costly one: the most straightforward way to realize this solution is to use a low-cost, capable of performing the desired procedures, smartphone starting from approximately $60.

Installation and cabling: Installation and cabling could be a very challenging task, since most venues follow strict policies regarding physical interventions in their premises. Despite the fact that RFID installations seems the easiest approach, since it does not require many physical changes of the exhibition space, nevertheless it requires many changes to the building, since it demands the use of RFID antenna grids. On the other hand, the use of complex installation equipment for the exhibits (memory-rich and agent-based) does not require any change to the actual venue building but a large device must accompany each exhibit, since individual power sources are needed, leading to extensive cabling in the exhibition space.

Compatibility with existing software: When dealing with software installed in museums we need to take under consideration the fact that existing narrative generation systems for museums (e.g., [5, 30]) assume that a central database holding the exhibit information exists. The software to be installed and used for the implementation of exhiSTORY system is related to the actual system architecture selected. In case of elementary implementation, the existing infrastructure of a museum could be sufficient for making the system work. In every other occasion, though, there is strong need for installation of extra software as a centralized database needs to be used. Furthermore, the infrastructure has to be in-line with the security implementation which may require software changes to the server so as to be able to recognize the objects' certificates. Finally, the agent-based approach requires advanced software installed on objects' end devices that will manage the interaction between them and the users.

Security: While all implementation options can provide high levels of security, as described in Sect. 5.3.3, the two RFID-based solutions require only minimal knowledge and expertise from the museum staff, while the other two options necessitate appropriate expertise from the museum staff (configuration of wireless LANs and knowledge on public key cryptography), which has to be done once though.

Final selection: According to the discussion and analysis on the different approaches, it seems that if a museum is able to address the cost and in-place changes, the most preferable implementations is either memory-rich or agent-based, as they are much more advanced than the RFID solutions. Furthermore, comparing the final two candidates we are leading to the result according to the capability of utilizing existing software versus building an autonomous agent system. Despite that the second option is more flexible and advanced, if there is the option of utilizing already existing systems, the memory-rich implementation should be preferable.

5.4   System Architecture

The main purpose of the described system is to enhance the experience that a visitor can acquire in an exhibition place. This is achieved by altering the exhibits and the place setup in order to comply with every aspect of the system described. The architecture that is described in Sect. 5.8 is essential in order to give life to the set of smart exhibits. It includes the "smart space," the knowledge base and a set of intelligent modules.

5.4.1   The Smart Space

Mobility of the exhibits is a core concept in exhiSTORY, meaning that the smart space is a dynamically defined area that is able to identify changes in location, movements, additions, or subtractions of objects. In the cases of solutions with RFID, the smart space includes a dense grid of readers, while in other solutions a network of Wi-Fi access points is required to support the space. The precision of estimating the location should not be very high. It is required though to have a rough estimate of the position. Moreover, what is actually required is to be able to have the knowledge of the proximity between objects and information of the order of object viewing, as discussed in detail in Sect. 5.3.2. The Wi-Fi network is required in every solution as it is the medium through which the system is able to communicate with the visitors, so it should be such that can cover every space, or at least the spaces that are very close to the exhibits.

5.4.2   The Knowledge Base

As already mentioned in the description of the exhiSTORY system, we should be able to have deep knowledge of the data that accompany the objects, as well as information about the visitors (user profile) in order to be able to provide an advanced user experience. In this manner, the knowledge base of the system includes the following:

- *The museum's context.* The context consists of information about the style of every exhibition, including style, related and incongruous topics, and ways of presentation.
- *The museum semantics.* Semantics is curated information that is related to each institute's interest. This kind of information is by far more reliable than knowledge acquired from any other sources—Internet or information within the smart exhibits.
- *The museum media.* Multimedia that is available to the museum and can be used in the presentation of stories and can or cannot be used under the current institute setup.
- *Museum map information.* This information includes a map of the layout accompanied with metadata comprising the Wi-Fi Access Point or/and RFID reader positioning.

The aforementioned knowledge base items exist when the system is set up and is ready to be launched. These data are completed with information acquired during system execution and are related to the user profiling data.

These form the actual database of the system and should be present in any system installation in order to be able to create links between the vast amount of data. The connection between information that leads to the creation of stories for the maximization of the visitors' experience is done by the intelligent components.

5.4.3   The Intelligent Modules

Detailed definition of the intelligent modules remains out of scope of this paper as our main intention is to describe the system architecture that leads to exhiSTORY forming smart exhibitions. As illustrated from Fig. 5.8 that presents the system architecture, the intelligent modules include: the exhibit tracker, the semantics engine, the story finder and maker, the media engine, and the user profiling. They are all defined as part of the main component of the service that is executed within the scope of the exhiSTORY system as they are highly interconnected. The core software of the system is beyond the scope of this manuscript which focuses on the construction of the sensors' architecture.

The *exhibit tracker* is the simplest of the modules. By using either the grid of RFID sensors, the position of the Wi-Fi access points or the geolocation unit of the object it is able to estimate the exact location of the exhibit. *Media engine* is responsible for enriching the systems' outcomes with multimedia such as text, video, audio, images, animation, or 3D objects. By locating the semantic interconnection of the objects to the system generated story metadata it is able to provide relevant media to be selected for each generated story.

The semantics engine is in charge of collecting information about the exhibits and the topics that are related to them. Firstly, an amount of semantic information is directly related to the smart exhibit and more precisely the exhibits' "memory." Furthermore, additional information can be gathered directly from the knowledge base and from reliable online sources such as cultural or historical Wikis and encyclopedias.

Relying on the "x degrees of semantic separation" the *story finder and maker* analyzes the context of each exhibit and tries to locate bonds between the objects within the semantics. Detailed description of the procedure of this system is presented in [3]. It is important to note that the system is in search of non-trivial connections between the objects as it is expected that the museums' curators should be able to identify the obvious and trite. After locating the interconnection between the objects, the part of this module that is responsible for creating the stories formulates the scenario. The scenario is based on the place of the exhibit the sequence of objects that a visitor will see and the content that is related to the semantics.

Finally, the system is furthermore enhanced with the user profiling module which is responsible for collecting implicit and explicit information about user profiles whenever possible. By doing this the system is able to apply personalization techniques to the creation of stories tailor-made to the profile of each user.

After available stories have been ranked, the ones attaining the highest scores are suggested to the user to choose from.

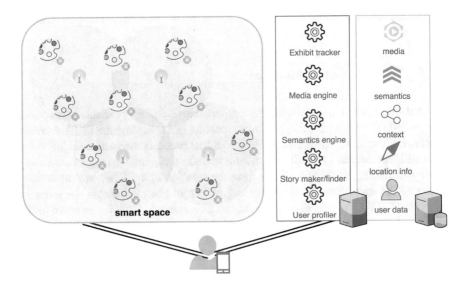

Fig. 5.8 The exhiSTORY architecture

Fig. 5.9 The exhiSTORY system in operation

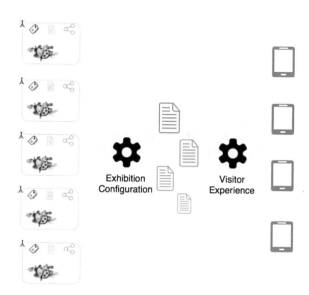

5.5 The exhiSTORY System in Operation

The operation of the exhiSTORY system (Fig. 5.9) is separated into two processes: the exhibition configuration and the visitor experience mode. Although the two processes are distinct, and can be associated to the offline and online modes of operation of most conventional systems, it is worth noting that they may be concurrent, as in exhiSTORY reconfiguration can take place at any moment.

The exhibition configuration process involves the geolocation sensors of the smart space, the exhibit tracker, the semantics and media engines, and the story finder and the story maker. It is the process that takes as input the list of smart exhibits and generates the detailed scenarios of the stories; its operation has been explained in Sect. 5.4.

The delivery of the visitor experience involves the Wi-Fi network, the user profiling module, and the story selector procedure. Additionally, since the memory-rich smart exhibits do not have incorporated display facilities, a mobile device is also necessary. A custom device could be created, but that is not a necessity as visitors' own mobile devices, such as smart phones and tablets, can be used, provided that the exhiSTORY application has been installed in them and is running.

The application can be downloaded and installed prior to the visit to the museum. Upon loading, the application prompts the user to connect via a Facebook account and/or to play some online games that are related to culture. While these steps are optional, they allow the system to compile an initial profile for the user, by sourcing relevant data from the user's Facebook profile or analyzing the user's behavior within the games to deduce personality traits of the user, such as MBTI dimensions (extraversion versus introversion, thinking versus feeling), etc. [2]. These profile data will then be exploited in the context of the personalization process. Additional sources of user profiling information, such as social network profiles [13] or web platform profiles [16], can be used.

When inside the museum, the application connects to the exhiSTORY system, transmitting to it user profile data and visitor history information, which is used by the story selector procedure to provide an ordered list the most prominent stories. When the user selects a story, the narration (and navigation) commences. Any mobile device equipped with a graphic screen, Wi-Fi connectivity and having the ability to execute custom applications network, such as a smartphone or a tablet, can be used as a user access device.

5.6 Discussion and Conclusions

It is inevitable that in the current state of museums the procedure of changing an exhibition by moving, altering, removing, or adding exhibits is a laborious one. Curators need to invest time and effort to set up a new exhibition, although the information presented is only a fraction of the available information for each exhibit. These all lead to the conclusion that it is very difficult to create media-rich exhibitions with an increased quality of experience for each visitor without the use of relevant technology. In this manner, we presented the exhiSTORY system, which allows smart exhibits to be automatically and dynamically organized into exhibitions. The exhiSTORY system delivers the narration of the exhibition through individual visitor devices, and considers user style, preferences, background, history, and so forth to present visitors with tailor-made experiences.

The exhiSTORY framework has been designed in part in the context of the Cross-Cult EU project, and will have its trial application in the implementation of one of the projects' pilots. That pilot will run at the Archaeological Museum of Tripolis in Greece, where we have found at least 7 different stories can be told by approximately 30 exhibits, depending on the configuration of the visit and the presentation. It will be fascinating to observe as the exhibition reconfigures itself when exhibits are added, removed, or simply rearranged in the museum's rooms. It will also be interesting to see how the exhibition evolves from day to day, based on social media trending topics, or how visitors in the same group are served with different contents. But, in all fairness, this trial will neither test nor demonstrate the full potential of the smart exhibit notion.

The concept of the exhiSTORY system is the "smart place" that includes smart self-organized autonomous objects. In this notion, we are creating a new research path with a large amount of challenges to be faced. We analyzed a number of alternatives that can be used in the technical part and described the solutions for each of the cases in order to produce a complete system, according to each museum's peculiarities. In these cases, we need to examine the interconnection of data in different museums as well as the cost analysis of the seamless and agent-driven implementations that seem to have large cost especially for museums with large numbers of objects. Solutions including novel sensors could be also investigated as well as custom devices based on electronic prototyping platforms like Arduino,[2] BeagleBone,[3] or Raspberry Pi.[4]

Acknowledgements This work has been partially funded by the project CrossCult: "Empowering reuse of digital cultural heritage in context-aware crosscuts of European history," funded by the European Union's Horizon 2020 research and innovation program, Grant#693150. This work has been partially supported by COST Action IC1302: Semantic keyword-based search on structured data sources (KEYSTONE).

References

1. Antoniou A, Lykourentzou I, Rompa J, Tobias E, Lepouras G, Vassilakis C, Naudet Y (2014) User profiling: towards a Facebook game that reveals cognitive style. In: Proceedings of the second international conference on games and learning alliance. LNCS, vol 8605. Springer International Publishing, pp 349–353
2. Bampatzia S, Bourlakos I, Antoniou A, Vassilakis C, Lepouras G, Wallace M (2016) Serious games: valuable tools for cultural heritage. In: Games and learning alliance conference (GALA 2016), December 5–8. The Netherlands, Utrecht
3. Bampatzia S, Bravo-Quezada OG, Antoniou A, Lopez Nores M, Wallace M, Lepouras G, Vasilakis C (2016) The use of semantics in the CrossCult H2020 project. In: 2nd international KEYSTONE conference (IKC 2016), Cluj-Napoca Romania, 8–9 September
4. Brachmann F (2012) A comparative analysis of standardized technologies for providing indoor geolocation functionality. In: Proceedings of the 13th IEEE international symposium on computational intelligence and informatics, pp 127–130
5. Broadbent J, Marti P (1997) Location aware interactive guides: usability issues. In: Proceedings of ICHIM 97, Paris, September 1997. https://pdfs.semanticscholar.org/2334/91cb0441914a3874d21d0258d98bc29c28b5.pdf. Accessed 28 June 2017
6. Bröring A, Datta SK, Bonnet C (2016) A categorization of discovery technologies for the internet of things. In: Proceedings of the 6th international conference on the internet of things (IoT'16). ACM, New York, NY, USA, pp 131–139. https://doi.org/10.1145/2991561.2991570
7. Cataloguing Cultural Objects Commons (2017) What is CCO? http://cco.vrafoundation.org/index.php/aboutindex/. Accessed 3 February 2017
8. Chianese A, Piccialli F, Valente I (2015) Smart environments and cultural heritage: a novel approach to create intelligent cultural spaces. J Locat Based Serv 9(3):209–234
9. Chianese A, Piccialli F (2014) Designing a smart museum: when cultural heritage joins IoT. In: Proceedings of the eighth international conference on next generation mobile applications, services and technologies. https://doi.org/10.1109/NGMAST.2014.21
10. Collections Trust (2011) The SPECTRUM Standard, v4.0. http://collectionstrust.org.uk/resource/the-spectrum-standard-v4-0/. Accessed 3 February 2017
11. de Freitas AA, Nebeling M, Chen XA, Yang J, Ranithangam ASKK, Dey AK (2016) Snap-to-it: a user-inspired platform for opportunistic device interactions. In: Proceedings of the 2016 CHI conference on human factors in computing systems (CHI'16). ACM, New York, NY, USA, pp 5909–5920. https://doi.org/10.1145/2858036.2858177
12. Diana ad Callisto. https://www.nationalgallery.org.uk/paintings/titian-diana-and-callisto. Accessed 18 August 2017

---

[2] Arduino: https://www.arduino.cc/.

[3] BeagleBord: http://beagleboard.org/.

[4] Raspberry Pi: https://www.raspberrypi.org/.

13. Facebook (2017) Facebook interest targeting. https://www.facebook.com/help/188888021162119. Accessed 28 June 2017
14. FlashAir Config, Network mode. https://flashair-developers.com/en/documents/api/config/#APPMODE. Accessed 28 June 2017
15. Gabrielli F, Marti P, Petroni L (1999) The environment as interface. In: Proceedings of the i3 annual conference: community of the future, October 20–22, Siena, pp 44–47
16. Google ads settings. https://www.google.com/settings/u/0/ads/authenticated. Accessed 28 June 2017 (prior login to a Google account is required)
17. Hashemi SH, Hupperetz W, Kamps J, van der Vaart M (2016) Effects of position and time bias on understanding onsite users' behavior. In: Proceedings of the 2016 ACM on conference on human information interaction and retrieval (CHIIR'16). ACM, New York, NY, USA, pp 277–280. http://dx.doi.org/10.1145/2854946.2855004
18. Hatala M, Wakkary R (2005) Ontology-based user modeling in an augmented audio reality system for museums. User Model User-Adapt Interact 15:339–380. https://doi.org/10.1007/s11257-005-2304-5
19. Hatthasin U, Vibhatavanij K, Worasawate D (2007) One base station approach for indoor geolocation system using RFID. In: Proceedings of Asia-Pacific microwave conference. https://doi.org/10.1109/APMC.2007.4554642
20. Houlberg Rung M (2016) Thoughts about an exhibition—using materiality and sensuousness as a narrative device. https://goo.gl/rpjjky. Accessed 28 January 2017
21. Hudson-Smith A, Gray S, Ross C, Barthel R, de Jode M, Warwick C, Terras M (2012) Experiments with the internet of things in museum space: QRator. In: Proceedings of the 2012 ACM conference on ubiquitous computing (UbiComp'12). ACM, New York, NY, USA, pp 1183–1184. http://dx.doi.org/10.1145/2370216.2370469
22. Ivan K, Steven L (1991) Exhibiting cultures: the poetics and politics of museum display. Smithsonian Institution Press, Washington
23. Klingemann M, Doury S (2016) X degrees of separation: the hidden paths through culture. Google Arts & Culture Experiments. https://artsexperiments.withgoogle.com/xdegrees. Accessed 5 February 2017
24. Lord GD, Lord B (2009) The manual of museum management. AltaMira Press, Lanham, MD
25. Lord B, Lord GD, Nicks J, Britain G (1989) The cost of collecting: collection management in UK Museums. H.M.S.O., London
26. Lykourentzou I, Claude X, Naudet Y, Tobias E, Antoniou A, Lepouras G, Vassilakis C (2013) Improving museum visitors' quality of experience through intelligent recommendations: a visiting style-based approach. In: Intelligent environments (workshops), 2013
27. Marti P, Gabrielli F, Pucci F (2001) Situated interaction in art. Pers Ubiquit Comput 5:71–74
28. Mok E, Retscher G (2007) Location determination using WiFi fingerprinting versus WiFi trilateration. J Locat Based Serv 1(2)
29. Naudet Y, Lykourentzou I, Tobias E, Antoniou A, Rompa J, Lepouras G (2013) Gaming and cognitive profiles for recommendations in museums. In: Proceedings of the 8th international workshop on semantic and social media adaptation and personalization (SMAP)
30. Petrelli D, Not E, Zancanaro M (1999) Getting engaged and getting tired: what is in a museum experience. In: Proceedings of the workshop on 'attitude, personality and emotions in user-adapted interaction' held in conjunction with UM'99, Banff, 23 June 1999
31. RFID Journal (2017). How much information can an RFID tag store? https://www.rfidjournal.com/faq/show?66. Accessed 4 February 2017
32. Rocchi C, Stock O, Zancanaro M, Kruppa M, Kruger A (2004) The museum visit: generating seamless personalized presentations on multiple devices. In: Proceedings of the 9th international conference on intelligent user interfaces, Funchal, Madeira, Portugal, January 2004, IUI'04. ACM, New York, NY, pp 316–318
33. Speed C, Shingleton D (2012) Take me I'm yours: mimicking object agency. In: Proceedings of the 2012 ACM conference on ubiquitous computing (UbiComp'12). ACM, New York, NY, USA, pp 1167–1170. http://dx.doi.org/10.1145/2370216.2370465

34. Stock O, Zancanaro M, Busetta P, Callaway C, Krüger A, Kruppa M, Rocchi C (2007) Adaptive, intelligent presentation of information for the museum visitor in PEACH. User Model User-Adapt Interact 17(3):257–304
35. Thaggert M (2010) Images of Black modernism: verbal and visual strategies of the Harlem Renaissance. University of Massachusetts Press, Amherst
36. The Arnolfini Portrait. http://photodentro.edu.gr/lor/retrieve/44427/15arnol.jpg. Accessed 18 August 2017
37. The NMC Horizon Project, NMC Horizon Report Museum Edition. http://files.eric.ed.gov/fulltext/ED559359.pdf
38. Ting SL, Kwok SK, Tsang AHC, Ho GTS (2011) The study on using passive RFID tags for indoor positioning. Int J Eng Bus Manag 3(1). https://doi.org/10.5772/45678
39. Toshiba (2017). FLASHAIR™. http://www.toshiba-memory.com/cms/en/products/wireless-sd-cards/FlashAir/. Accessed 4 February 2017
40. Viswanathan V, Krishnamurthi I (2017) Finding relevant semantic association paths through user-specific intermediate entities. Human-centric Comput Inf Sci 2. Article Number 9(2012). https://hcis-journal.springeropen.com/track/pdf/10.1186/2192-1962-2-9?site=hcis-journal.springeropen.com. Accessed 28 June 2017
41. Wecker AJ, Lanir J, Kuflik T, Stock O (2015) Where to go and how to get there: guidelines for indoor landmark-based navigation in a museum context. In: Proceedings of the 17th international conference on human-computer interaction with mobile devices and services adjunct (MobileHCI'15). ACM, New York, NY, USA, pp 789–796. http://dx.doi.org/10.1145/2786567.2793702

Chapter 6
IoT Cloud Security Design Patterns

Bogdan-Cosmin Chifor, Ștefan-Ciprian Arseni, and Ion Bica

Abstract Internet-of-Things (IoT) devices are getting increasingly popular, becoming a core element for the next generations of informational architectures: Smart City, Smart Factory, Smart Home, Smart Health-Care, and many others. IoT systems are mainly comprised of embedded devices with limited computing capabilities while having a Cloud component that processes the data and delivers it to the end-users. Many of the IoT applications handle sensitive user data, thus needing an appropriate security solution. This chapter analyzes the IoT design patterns from multiple perspectives (infrastructure, interaction, application programming, information model, deployment scenarios) and identifies the security solutions required for each scenario. The IoT Cloud-enabled devices paradigm is becoming an industry standard, being supported by the open-source community and also by multiple service providers and device manufacturers, each with its own custom solution. Cloud-integrated IoT devices facilitate the design of microservice architectures, where each sensing device exposes a service managed through an application programming interface (API). IoT security solutions are not limited to remote embedded devices and Cloud computing boundaries, being integrated with user-managed devices and software applications. One of the most encountered IoT design patterns in the Smart Home scenario consists of smartphone-controlled devices. Regarding this security design pattern, the capabilities of integrating Android smartphones, IoT devices, and Cloud platforms are presented in this chapter. The most critical IoT architecture element is the network infrastructure along with the IoT network protocols. The gateway-centric network is a solution adopted by the most important hardware manufacturers, almost every one of them offering devices with built-in IoT data processing and transport features. Because most embedded networks exchange

B.-C. Chifor · Ș.-C. Arseni · I. Bica (✉)
"Ferdinand I" Military Technical Academy, 39-49 George Coșbuc Blvd., Sector 5, 050141 Bucharest, Romania
e-mail: ion.bica@mta.ro

B.-C. Chifor
e-mail: bogdan.chifor@mta.ro

Ș.-C. Arseni
e-mail: stefan.arseni@mta.ro

© Springer Nature Switzerland AG 2021
F. Pop and G. Neagu (eds.), *Big Data Platforms and Applications*,
Computer Communications and Networks,
https://doi.org/10.1007/978-3-030-38836-2_6

113

sensing information in a publish–subscribe manner, they are suited for data-centric network paradigm. This chapter evaluates the security issues raised by data-centric elements deployed in IoT networks.

Keywords Internet-of-Things · Cloud computing · Security architectures · Design patterns

6.1 Introduction

Internet-of-Things (IoT) is an emerging subject that enables various applications in different areas with great impact in our society, such as Smart City, Smart Home, Smart Transportation, health-care industry, and many others. The IoT field brings new technology perspectives by connecting embedded devices to the Internet, being considered one of the engines for the fourth industrial revolution. Even though it is considered that the IoT domain is comprised of network-connected embedded devices, it also encompasses various technologies like Cloud and Edge services, that process data provided by the IoT devices in a near real-time manner. Multiple studies predict an important number of IoT devices to be deployed in the near future [1], thus stressing the great impact that IoT applications will have [2, 3]. Security is a critical field in this context [4], being the factor that could accelerate the large-scale adoption of this technology [5]. In an IoT scenario, a security vulnerability could cause more losses than disclosing the user's private data, while in a context like the cyber-physical systems (CPS) it could be a threat even for human lives [6]. An IoT application is a complex system that contains a heterogeneous suite of hardware devices: endpoint embedded modules, gateways, Edge and Cloud servers. The deployment of these hardware elements enables a multi-layer IoT application structure with an endpoint device, gateway, Edge and Cloud layers. Each IoT layer runs a different software stack, thus an IoT application security solution must be tailored for each computing layer.

The classic Cloud security threats [7] are translated into the IoT context and even augmented with new types of attacks, specific to resource-constrained embedded devices. The IoT systems need novel security protection mechanisms that address the requirements of each computing layer. The endpoint IoT layer is populated with resource-constrained devices, thus the security defense capabilities are limited: classic security mechanisms like resource-intensive encryption, network packet filtering or security policy enforcement are not feasible for this context. Even though IoT applications are used in a multitude of scenarios, each one addressing a particular problem, these applications tend to follow a suite of architectural design patterns. These architectural design patterns ease the development of IoT solutions due to the fact that a series of software and deployment paradigms can be reused, thus reducing the development costs. From the security point of view, using architectural design patterns reduces considerably the time to market of an IoT solution, because an existing security solution can be employed with minimum effort. Also, by following a certain design pattern, the security context of an IoT deployment scenario can be evaluated easier: the mitigation process of certain threats along with the risk assessment model. In a software application, a design pattern denotes a modality to solve a commonly encountered issue, such as code modularization (e.g., model-view-controller architectural pattern) or single-responsibility modules. In a similar way, an IoT architectural design pattern encompasses common solutions for designing/developing and deploying an IoT solution. Taking into consideration the limited capabilities of endpoint IoT devices, various IoT design patterns involve a Cloud component that manages the entire IoT application, having system-wide visibility [8]. The design pattern concept is also presented in [9], as an enabler for the development of IoT systems following an open architecture that eases system integration and interoperability.

In a typical IoT application, a Cloud platform is a data consumer that processes the information acquired and delivered by the endpoint IoT nodes. After the data is processed, it can be published for the end-user in a specific presentation format (e.g., web or smartphone application) and subsequent IoT actuation commands may follow after the user input or after the IoT-generated information is processed. Regarding the IoT Cloud paradigm, an IoT application can follow two models: private (or on-premise) Cloud or a software-as-a-service (SaaS) Cloud platform, both models bringing advantages and drawbacks. In this context, a private Cloud does not necessarily mean a user-owned hardware, but a user-owned Cloud application that communicates with the remote IoT layers. An alternative to the private Cloud is the SaaS platform that could offer data processing, analytics, IoT actuation or machine learning services. Several Cloud providers, such as Amazon, Google, Microsoft or Bosch, are offering these types of services, each one with a custom IoT device management paradigm along with an associated API (Application Programming Interface). On one side, the advantage of a private Cloud application consists in the possibility of tailoring the designed Cloud services for the users' IoT application, while having the drawback of increased CAPEX (Capital expenditures) and OPEX (Operating expenses) associated with the Cloud hardware and infrastructure. On the other side, the advantage of a SaaS Cloud is the security and scalability that comes out of the box, while having the drawback of being vendor dependent, thus using proprietary Cloud APIs. From the security point of view, another disadvantage of using IoT SaaS is disclosing the IoT generated private data to the Cloud platform, in the context of data processing/machine learning services. A Cloud platform is a module employed by an IoT application due to its capacity to offload the endpoint network from complex data processing. This feature raises an interoperability issue between different IoT Cloud providers, taking into account the lack of standardization in the IoT field. Thus, transitioning from an IoT Cloud provider to another one requires development and testing resources, the lack of standardization also affecting a multi-cloud IoT application scenario, where an endpoint network employs back-end services from different providers. The lack of standardization also impacts the IoT security field because a common set of out of the box security designs addressing a suite of IoT scenarios, along with the associated security protocols (e.g., authentication, authorization, audit, etc.), is missing.

Cloud integrated IoT devices facilitate the design of microservice architectures, where each sensing device exposes a service managed through an API. Such a system raises a multitude of security issues like authorization and reputation because the information is handled by untrusted entities that process sensitive data or provide microservice composition. The microservice architecture is a highly adopted design paradigm in the Cloud world because it allows the application module segmentation. The software module segmentation brings multiple benefits, such as software modularity (and ease of maintenance), a small code base, and a simple software module replacement/upgrade. In the microservice paradigm, every software module must implement a well-defined interface or a software contract. In an IoT system, that is by definition a heterogeneous one, the microservice paradigm is a modality to integrate multiple software modules: Cloud components and embedded devices. From the security point of view, employing microservices facilitates the security tasks delegation: for instance, a microservice may handle only the authentication process. Thus, a security microservice having a single responsibility can have a small code base that is easier to be evaluated, therefore enabling a trusted computing base (TCB). Regarding the costs associated with the IoT application maintenance and module replacement, an IoT microservice represents an advantage because the developers can easily migrate from an IoT Cloud service to another one if the back-end service functionality is abstracted by a well-defined and vendor-independent interface.

IoT security solutions are not limited to remote embedded devices and Cloud computing boundaries, being integrated with user-managed devices and software applications. One of the most encountered IoT design pattern, in a Smart Home scenario, consists of smartphone-controlled devices. This pattern uses a smartphone application as a security anchor that controls, authenticates, and authorizes certain device actions. Considering the hardware and software architecture of smartphones, this solution is highly adopted because it provides both user-to-device (by means of biometrics) and device-to-user authentication.

Taking into consideration the increased number of IoT devices, one of the most important design patterns is the device centralized management, by using master data management solutions (MDM) and IoT Software Defined Networks (SDN), for controlling and updating the IoT device fleet. These applications require authentication, attestation, and confidentiality solutions while providing separate control and data planes. The most critical IoT architecture element is the network infrastructure along with the IoT network protocols. Taking into consideration the IoT Cloud devices paradigm, the network infrastructure has a gateway that links the devices with the Cloud components and acts as a translator from lightweight communication protocols. The IoT gateway acts as a trusted security element while offloading the endpoint devices from certain computing-intensive tasks. The gateway-centric network is a solution adopted by the most important hardware manufacturers, almost every one of them offering devices with built-in IoT data processing and transport features. IoT communication protocols are designed to be lightweight and suitable for resource-constrained devices, while the security component is left to be managed by each application, thus being an element that requires attention. Because most embedded networks exchange sensing information in a publish–subscribe manner, they are suited for the data-centric network paradigm. IoT endpoint software stack runs on different systems, ranging from bare-metal to custom Linux distributions. Such devices require security frameworks with low code and memory footprint, equipped with lightweight cryptography and DTLS (Datagram Transport Layer Security) capabilities.

In this chapter, we explore possible solutions for the main security issues raised by data-centric elements deployed in IoT networks, by analyzing the characteristics of various design patterns that can be used in different layers of an IoT system. We first present the main elements of an IoT system, then we continue addressing the main topic of the chapter by assessing several design patterns in two sections: one in which we target the network layer, seen as the backbone for every IoT system and complementary for the Cloud layer, and another one in which we present design patterns used for implementing Cloud platforms, the central point linking users and the smart environment.

6.2 Design of IoT Architecture Layers

As previously mentioned, an IoT system is a complex architecture composed of several modules: the communication network, the endpoint devices, the local gateway, the Edge and Cloud modules, and the management platform. From the security point of view, the IoT system segmentation in a suite of sub-modules brings many benefits, mainly by allowing a security solution to address the issues of a particular IoT layer. Due to the diversity of IoT applications, designing a generic security solution is not feasible. From the functional point of view, an IoT system has the following characteristics:

- *Mobility*—from the communication protocols' perspective, mobility is equivalent to a dynamic and lossy network, where a device can change its location and IP address or establish ad-hoc connections with potentially untrusted devices.
- *Availability*—one of the main features of an IoT application is the availability and the system capability to process data and respond in a near real-time manner.
- *Scalability*—another important feature of an IoT application refers to its capacity of sustaining a network extension or integration/discovery of new services.
- *Administration*—given the complexity that IoT systems usually have, the main requirement is to have a friendly management framework that will allow an administrator to easily configure and collect data from the IoT devices in a secure and reliable manner.
- *Interoperability*—the complexity of an IoT application implies also a heterogeneous pool of IoT devices, thus it is important for an IoT application to integrate a system for translating messages through various protocols.

The modules comprising a general architecture of an IoT system can also be grouped in specific functional layers, as presented in Fig. 6.1, namely IoT Nodes,

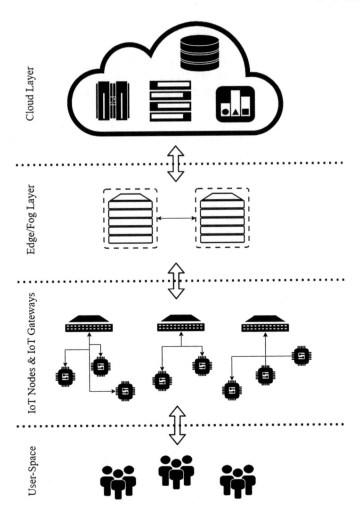

Fig. 6.1 General architecture of an IoT system

Fog/Edge and Cloud Layers. As presented, the main element of this system is represented by the endpoint IoT nodes which are interconnected by means of the local
IoT network. The central element of this architecture is represented by the Cloud
module that is in charge of managing the IoT nodes as a single system. The local
IoT layer is managed by means of local gateways, which have both functional tasks
(e.g., protocol translation) and security tasks (e.g., authenticating and authorizing the
client IoT devices). The Cloud application is a software module that has a system-
wide visibility and executes complex data processing, connects IoT sub-systems
from different locations and allows the IoT system management.

Some IoT architectures also have an intermediate layer, located between the
local gateway and the Cloud: the Fog or Edge layer. This layer can be considered a

simplified Cloud module, allowing fast data access. Another layer of an IoT architecture is represented by the software applications designed for user interaction, deployed in the local IoT network or in an adjacent network. The user interaction layer generally consists of smartphone or web applications that are consuming IoT generated data and send actuation/management commands to IoT devices. Regarding the security aspects of the user interaction IoT layer, the main requirement is the user-experience, thus any user relation with IoT authentication or authorization solutions must meet a trade-off between the desired security level and the ease of use. In an IoT architecture, the gateway has an important role, from both the functional and the security point of view, being the element that makes a connection between the IoT local network and the upstream network segment. The gateway is the element that translates lightweight communication protocols, used in the downstream network segment, to classic communication protocols (e.g., HTTPS—Hypertext Transfer Protocol Secure), used in the upstream network segment.

6.2.1 Security Aspects

When discussing the security aspects of an IoT environment, the need for network segmentation should be taken into consideration, given the various types of devices that are interconnected, each one of these types having different access rights and security levels assigned to them. Due to this, the design process of an IoT network topology is influenced by the communication protocol used as a backbone for the information flows inside the IoT system. Therefore, an IoT system can be included in one of the multiple categories of network designs, such as client-server or publish–subscribe. Yet, recent years have brought an increase in IoT devices functionalities and inclusion in multiple layers of society, which in return lead to an integration of multiple communication protocols in local networks and sometimes even to overlapping of network topologies in the same IoT environment.

6.2.1.1 Client–Server Paradigm

By analyzing the majority of IoT networks topologies from the perspective of integrated communication protocols used at the level of local networks, two of the commonly used protocols are CoAP (Constrained Application Protocol) and MQTT (Message Queuing Telemetry Transport), each one implementing a different paradigm: client–server for CoAP and publish–subscribe for MQTT. Focusing on CoAP, a device can be a *Client* (sends requests to a CoAP server through unicast or multicast packets), a *Server* (answers to the requests received from clients) or a *Proxy* (acts as a relay between clients and servers). The client–server paradigm requires a link between a client and a server and from a security point of view, the way this link is constructed requires total trust between all the devices that contribute to the formation of this communication path (especially a proxy), mainly because of the susceptibility to be vulnerable to MiTM (Man-in-The-Middle) attacks. Moreover, the link needs to be secured using a cryptographic algorithm that both the client and the server need to know and agree upon.

6.2.1.2 Publish–Subscribe Paradigm

Moving on to the second paradigm, publish–subscribe, this requires the existence of a broker, a specially configured device that can redirect messages from a sender to a list of receivers. Still, this broker is not mandatory, given that there are some communication protocols in this paradigm that use multicast packets to send information, such as the DDS (Data Distribution Service) protocol. The publish–subscribe architecture is constructed based on the following roles that IoT devices can have: Subscriber (or data consumer, is the end device that connects to the information updates of other devices and consumes the received data to ensure its functionality), Publisher (or data sender, is the device that publishes information on a specific topic that Subscribers consume) or Broker (message distributor that facilitates the communication between a Publisher and a list of Subscribers).

From a security perspective, this second paradigm brings many advantages but has also some disadvantages, mostly introduced by the Broker, representing a *single point of failure*, that once it is compromised, the entire IoT system is exposed, giving an attacker access to all of the devices that are connected in that system. Other concerns that may arise from the usage of an unique message dispatcher are oriented toward the network availability, given that in case of failure the entire information flow inside the IoT system is halted. Despite these disadvantages, a central broker can ease the management and control of all devices that are connected to an IoT network, mainly by specifying a central element that can provide mechanisms to authenticate devices and, if needed, impose certain access control rules. Another advantage is reducing the network complexity, this central element transforming the network in a star topology, where each device needs only to establish a trust relationship with the broker to have access to the entire network.

6.2.1.3 IoT Systems Segmentation

No matter what communication paradigm is implemented in an IoT system, all of them rely on the same basic segmentation: the base segment (gateways) collects data from devices and delivers it to an upper layer that will, in general, perform a data filtration and aggregation process. This upper segment can consist directly of a Cloud system or an intermediate layer, like an *Edge* layer. In case of an Edge layer, the components that can be included here are mainly platforms offering low latency data transport and providing different services, such as data processing or storage. As highlighted in [10], Edge and Cloud modules can be more easily distinguished if the following characteristics are taken into consideration when analyzing these types of modules:

- *Localization near the local IoT network*—offers low latency by processing data near the location, thus reducing the time and bandwidth needed to transfer data directly to Cloud services/modules.
- *Mobility*—being directly connected to a local IoT platform that can be mobile, it needs to provide support for mobility protocols, such as LISP (Locator/ID Separation Protocol).
- *The ability to process real-time data*—critical IoT systems often need real-time or near real-time responses after data collection. Given that a Cloud platform introduces latency not only in data transmission but also in resource allocation, data processing and storing, an Edge module can prove to be a more suitable solution.
- *Interoperability*—sometimes, data collected from IoT nodes could be processed by different services, exposed by Cloud modules. In this case, Edge modules need to provide the means required to connect to multiple Cloud service providers, therefore a multitude of communication protocols need to be integrated into these modules.
- *Provide support for data analysis*—similar to Cloud modules, Edge elements need to have the functionality of data analytics. The low latency response can be sufficient for certain tasks and sometimes even substitute a Cloud service.

IoT platforms are often defined as a multitude of technologies that enable not only the grouping of IoT elements but also the extraction, analysis, and presentation of data to the final user, in a transparent and effortless manner. As previously mentioned, IoT chain value can be simplified to the following fundamental layers: *Layer 1*—sensors and smart devices that produce data; *Layer 2*—Gateways that group and ensure network connectivity between Layer 1 elements; *Layer 3*—solutions that process, analyze, and format data gathered from Layer 1 elements; *Layer 4*—services or modules that present formatted data to the user and provide means of managing Layer 1 elements. Based on this layer classification, an IoT platform can be enclosed at the boundary between Layers 3 and 4 and can be defined as *middleware* between IoT devices and users, mainly because it provides the services and functionalities that enable the latter to control the former and analyze data extracted from them. Given the continuous evolution of technology, these IoT platforms have migrated mostly in Cloud, either proprietary or public, thus becoming a *PaaS* (*Platform-as-a-Service*) and bringing numerous benefits, such as scalability, ease of integrating various communication protocols, user and device protection through access control or other security mechanisms, global access to data through web services.

6.3 IoT Network Design Patterns

Networking protocols are considered the core of an IoT architecture, being the infrastructure that allows data passing between endpoint IoT nodes and the upstream network segment composed of gateways, Edge or Cloud computing nodes. Taking into consideration the constrained nature of embedded IoT devices and the limited security defense capabilities, the networking infrastructure is the most important security attack surface. Along with the on-board sensors, the network card is the main input channel of an IoT device, thus being targeted by attackers that try to manipulate the input data. In terms of complexity, the IoT network protocols can be analyzed by taking into consideration the IoT application layer: endpoint network, Fog/Edge, Cloud. The endpoint network employs lightweight communication protocols (e.g., MQTT, CoAP, DDS), while the upper layer IoT application modules employ traditional communication protocols like HTTPS (e.g., REST services).

6.3.1   Security of IoT Networks

From the security point of view, the communication protocols on the endpoint network need novel security solutions due to the multitude of attacks that IoT devices are vulnerable to. In an IoT network, the lightweight communication protocols are simplified versions of traditional protocols or newly designed ones, customized for resource-constrained devices. Taking into account this observation, the existing network protocols attack surface is extended with vulnerabilities associated with lightweight communication protocols, such as DoS, buffer exhaustion or even intensive protocol use. This security context becomes even more difficult because the defense capabilities of a resource-constrained IoT device is extremely limited: complex firewalls, IDS/IPS or antivirus solutions are not suitable. The lightweight communication protocols span over the entire networking stack, starting from Layer 2 protocols and ending with application protocols. Even though the IoT network topologies can be very diverse, these architectures tend to follow certain design patterns that ease the development of security solutions that target the IoT communication protocols. Due to the diversity of the IoT network applications, it is difficult to design a general security solution. By addressing the requirements of a certain IoT network design pattern, a security solution covers various deployment scenarios, thus being used by various applications without requiring major modifications. This section analyzes several network security designs in order to ease the evaluation of an IoT network security solution. As presented in [11], an IoT network stack is vulnerable to the following category of attacks:

- Physical layer attacks

  - *Jamming attack*—interference with the transmitted data on a certain frequency.
  - *Tampering attack*—an attacker gains physical access to the IoT device and manipulates the communication channel.

- Link layer attacks

  - *Exhaustion attack*—the targeted IoT device battery is exhausted due to the big number of processed malicious packets.

- *Retransmission attack*—the malicious packets are transmitted on the same frequency as the legitimate packets, thus causing packet retransmission.
- *Sybil attack*—the attacker impersonates a legitimate IoT node and injects malicious packets into the network in order to manipulate the communication between two malicious nodes.

- *Adaptation layer attacks*

  - these types of attacks target the fragmentation and reassembly packet process, for instance in the context of 6LoWPAN protocol.

- Network layer attacks

  - *Sinkhole attack*—legitimate network packets are captured by an attacker.
  - *Hello flood attacks*—malicious Hello packets are transmitted in order to drain the battery of the device.
  - *Blackhole attack*—a malicious node with routing capabilities captures the legitimate packets, thus disrupting the communication between legitimate nodes.
  - *Sybil attack*—a malicious node impersonates a legitimate node with routing capabilities and injects false network routes.
  - *Wormhole attack*—two malicious nodes advertise routes with better metrics and transmit the captured packets through a tunnel.
  - *Spoofing attack*—a malicious node advertises false routing information in order to disrupt the routing process.
  - *Neighbor discovery attack*—a malicious node publishes false network adjacency information in order to manipulate the communication.

- Transport layer attacks

  - *Flooding attack*—the malicious device opens multiple connections in order to drain the resources of the attacked device.
  - *Synchronization attack*—the legitimate node is forced to process multiple synchronization messages.

- *Application layer attacks*—the application layer attacks can differ, most of them falling into the DoS attack category. An attacker targets the application layer of an IoT architecture by exploiting proprietary characteristics of the application communication protocol.

6.3.2 Design Patterns for a Secure IoT Network

To mitigate part of the previously mentioned category of attacks, an IoT system must implement various security mechanisms that target authentication, authorization, confidentiality along with complementary measures, such as packet filtering, deep packet inspection, DoS mitigation, and others. Taking into consideration the fact that most of the IoT security attacks are executed in the network application protocol plane, the application layer needs security solutions and design patterns in order to handle the most encountered IoT security issues.

6.3.2.1 Central Broker Pattern

IoT networks are dominated by the publish–subscribe communication paradigm, thus the presence of a central broker is required. From the packet dispatching perspective, the broker relays the packets from the publishers to the subscribers, based on the message *topic*. From the security perspective, the broker represents a network entry point, handling sensitive operations like authentication and authorization. Moreover, due to its nature, the broker can handle other security-related operations like packet filtering or deep packet inspection, having quick access to the network packet data. In a general IoT network, the broker is considered a device with higher computational capabilities, in terms of CPU and memory, than a regular endpoint IoT node. This transforms the broker in a device that can offload certain network services, including security operations. Due to these capabilities, the broker is a security target in the local network, being targeted especially by DoS attacks. Even though the central broker pattern is applied mostly to publish–subscribe IoT networks, this model also appears in the client–server networks, where an intermediary device bridges the communication at the application level. Considering this, the central broker pattern can be applied to both MQTT/MQTT–SN and CoAP IoT lightweight protocols. For the MQTT and the MQTT–SN protocols, the broker is a mandatory element, being part of the protocol specifications. For the CoAP protocol, the central broker takes the form of a proxy element that processes the CoAP requests along with the associated CoAP responses. For all of these protocols, the central broker can be a policy enforcement point (PEP), being able to execute various security processes in order to ensure all the protocol parties (e.g., both the publisher and the subscribers) that the other part is genuine and sends legitimate data. The central broker pattern acts only in the local network and does not have implications in processing and sending the data to the upstream network segment. The central broker paradigm is presented in Fig. 6.2. The most important security tasks that apply to the central broker pattern are the following: authentication, authorization, security policy enforcement. For the central broker security tasks to be executed successfully, a trust relationship must be established between the broker and the IoT endpoints that are sending and receiving the protocol packets. The trust relationship is established in another phase and is usually handled by means of other design patterns. The authentication process is handled by using the following mechanisms:

- *Username/password*—in this situation the IoT device must be provisioned with a username and a password. The credentials-based authentication has two flavors: it can be executed as part of the communication protocol (e.g., MQTT password-based authentication) or it can be executed using a third party communication channel, thus being communication protocol agnostic. Both authentication flavors bring advantages and drawbacks: the in-band protocol password authentication

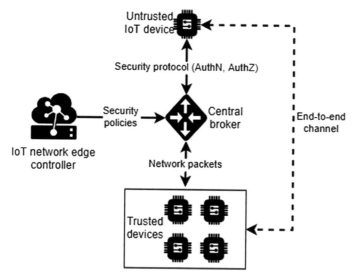

Fig. 6.2 Central broker paradigm

flavor requires no other third-party software for handling this process while having to be replicated for each employed protocol, while the out-of-band password authentication brings protocol independence and decouples the authentication process from the employed communication protocol, having the disadvantage of installing and maintaining another network service. The password is an application-specific field that can be populated with any security information, this being in line with the MQTT standard. For instance, the Google Cloud IoT Core uses a JWT (JSON Web Token) as password [12]. Thus, an IoT device can sign JWT data with a private key and prove its identity.

- *Symmetric cryptographickeys*—cryptographic pre-shared keys are a common authentication method used especially for resource-constrained IoT networks. This type of cryptographic key has the advantage of being used in lightweight symmetric or HMAC cryptographic algorithms. Depending on the IoT network security requirements, there can be a pre-shared key for a group of devices (or even the entire network), or there can be a unique key shared between the central broker and the IoT node.

- *Raw asymmetric cryptographickeys*—asymmetric cryptography is a feasible IoT authentication mechanism, allowing a device to use a public key as identity and to employ the private key (e.g., in a challenge signature process) to prove its identity. Although most of the asymmetric cryptographic algorithms are more computing-intensive than the symmetric ones, there are alternatives to the classical RSA algorithm, like ECC, that are suitable for a certain category of IoT devices.

- *Digital certificates*—although not widely used in the IoT endpoint networks, the digital certificates (X.509) are a classical way of authenticating a device.

Compared to the raw cryptographic keys authentication solution, digital certificates provide an easy way of linking the device identity to the public key. The overhead brought by the digital certificate authentication solution consists of the existence of a certificate authority (CA) that issues the certificates. For the central broker pattern, the CA and the broker roles can collude in order to simplify the overall security network design.

While the username/password authentication method is handled by an application protocol, the other authentication mechanisms may be handled by a lower layer protocol (for instance a transport protocol), thus being application protocol agnostic. For instance, raw asymmetric cryptographic keys and digital certificates can be used as an authentication method for the TLS and DTLS protocols. TLS and DTLS are transport protocols, thus can be used with both MQTT/MQTT-SN and CoAP protocols. In the central broker pattern, each IoT node establishes an authenticated transport session with the broker, acting as a trusted third party in the network. Considering this, the central broker pattern brings scalability and security design simplicity, the end-to-end trust being delegated to the central broker. The password-based authentication method is considered weak compared to the ones based on cryptographic keys. The evaluation of password-based authentication methods is realized in the user context, where password complexity is an issue that must be taken into consideration. For an IoT device, the password authentication feasibility is given by the password management system (generation, renewal, etc.), this method being equivalent to the symmetric key (pre-shared) authentication method. In the security provision phase, the IoT device password can be generated by a central broker or it can be generated by the device and registered at a later time to the central broker. The key security element for the password authentication method is the requirement of a dynamically generated password, which is not stored directly into the IoT device binary application and the central broker configuration files. The MQTT protocol [13] handles the username/password data into the variable header of the protocol, in the MQTT CONNECT packet. The presence of the username and password in the MQTT protocol is signaled by a series of associated bit flags in the CONNECT packet. When the MQTT central broker receives a CONNECT packet with username and password, the deployed MQTT implementation can choose the modality of authenticating the IoT device, this being application-specific. The authentication method can rely on a local identity repository (e.g., a local database, storage file or operating system) or it can employ an external identity repository: LDAP or other similar identity directories. The MQTT specifications do not indicate any password protection mechanisms, thus the MQTT implementation must develop a password protection mechanism or it must use a secure transport layer that ensures the confidentiality between the client and the central broker. A password protection mechanism can be easily implemented by using a challenge-response protocol or any similar method that allows an IoT device to prove the password possession without sending the password in a cleartext form to the central broker. The password authentication mechanism can be used in collaboration with the server-side TLS protocol, thus being ensured the communication channel confidentiality, the server-side TLS also allowing the endpoint IoT device to authenticate the central broker. As mentioned in the MQTT specifications, the MQTT protocol is not symmetric, thus the standard does not provide an MQTT specific modality for the client to authenticate the central broker. The central broker authentication can be achieved by using a transport-specific mechanism—like TLS/DTLS, or by using an application-specific solution—a central broker authentication protocol that is transported as MQTT payload. In a similar fashion to the MQTT central broker authentication, CoAP does not have any protocol-specific authentication methods, delegating this task either to the lower transport protocol or to the upper application protocol. Taking into consideration the CoAP UDP-based transport, any authentication method applicable to the DTLS transport protocol can be used in the CoAP context. Moreover, the application can use a custom authentication protocol that can be transported in the CoAP message payload. Another security task executed using the central broker design pattern is the IoT device authorization. The authorization process employs the IoT device identity, obtained from the authentication phase, and determines the subset of operations that can be executed by the IoT device. Besides the IoT device identity, an authorization process can use, as authorization criteria, an extended information set, like network-related information—MAC address, IP address, transport information, etc. Most of the IoT application protocol authorization modules use only the device identity, without relying on contextual information. The central broker pattern generally uses two forms of authorization mechanism: a communication protocol encapsulated authorization mechanism or a protocol-agnostic authorization. The protocol encapsulated authorization solution has the drawback of having a protocol-wide visibility in terms of authorization criteria and permitted actions (limited set). For instance, an MQTT authorization module can use the device username as authorization criteria and a set of MQTT topics that can be used by the device to execute certain operations—such as read or write. The advantage of having a protocol-wide authorization module is the ease of administration and defining ACL rules using protocol-specific semantics (e.g., MQTT topics, MQTT methods, CoAP methods). A protocol-agnostic authorization module brings the advantage of defining a rich set of authorization criteria along with a single authorization module serving several communication sessions. Decoupling the authorization mechanism from the actual communication protocol software gives a modular solution, where the software maintenance/replacement process is eased.

6.3.2.2 Client–Server Pattern

Client–server is one of the most used communication models, this paradigm being adopted in the web context, where web servers are delivering HTTP resources to a wide variety of clients: personal computers, smartphones, and even embedded devices. In the IoT world, the client–server communication model is also widely employed, with the particularity of using lightweight communication protocols, given the computing limitations of IoT devices. The actors of a client–server communication design pattern are the following, as depicted in Fig. 6.3:

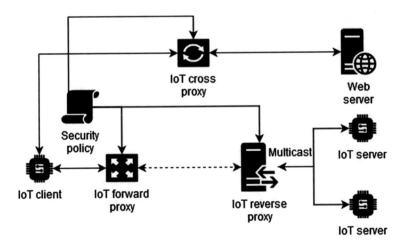

Fig. 6.3 Client-server communication

- The *client*—this is an IoT device or a management device that manipulates (via CRUD—Create/Read/Update/Delete operations) a resource on another IoT device.
- The *server*—this is a regular IoT device that exposes a resource to other devices in the network.
- The *proxy server*—this is an intermediate module that facilitates the communication between the client and the server. This module executes specific processes, such as security tasks or resource caching.

One of the most employed client–server communication protocols in the IoT networks is CoAP, a simplified version of the well-known HTTP protocol. CoAP uses UDP, which is a connectionless transport protocol and has a suite of IoT characteristics, such as binary encoding (low overhead), multicast support, caching or discovery. Being transported by an unreliable protocol, CoAP has an acknowledgment mechanism built in the application layer which consists of a message identifier and a message token. The previously mentioned values are used in matching the response with the request, the token having also a security purpose. The CoAP token is a randomly generated value (has up to 8 bytes) that can be used to detect replay attacks. The CoAP CON packet type requests a confirmation consisting of the CoAP ACK packet type (this packet type can also transport the CoAP response—a different packet type).

Another feature of CoAP is the multicast support, where an IoT device can send a CoAP request to a multicast address and receive multiple responses. Even though this feature brings a lot of advantages, like collecting data from multiple sensors by only sending a single CoAP message, from the security perspective this feature is an issue because it has to implement additional security mechanisms for group authentication. The CoAP RFC [14] mentions the impossibility of using the CoAP multicast along with DTLS transport, but recent advancements in DTLS present a multicast extension, like the one in [15].

Another particularity of multicast CoAP is the random period of time waited by the servers before sending the response: this mechanism avoids overloading the client IoT device with too many responses, thus causing a DoS attack. This feature is security sensitive because a malicious multicast server (or a group of servers) might bypass this implementation requirement and flood the client with responses, thus leading rapidly to a battery drain scenario. Another consequence of this attack could be the client's impossibility to process the other legitimate multicast responses. The CoAP multicast feature can also lead to a DoS attack triggered by a spoofing attack, where the malicious device can spoof the victim's address and send a multicast CoAP request, thus causing the victim to be flooded with unsolicited CoAP responses. Another type of attack enabled by the client–server model, especially by the multicast scenario, is the amplification attack, where an attacker can send a malicious CoAP request and generate larger CoAP responses that are redirected to a victim node. By exploiting the CoAP amplification vulnerability, an attacker with low resources can generate a large amount of traffic. Another component in the CoAP client–server design pattern is the proxy element that can have the following types:

- *Forward proxy*—executes CoAP requests for the client IoT devices. By using this module a series of security tasks can be executed, such as: authentication/authorization, response rate-limiting, etc.
- *Reverse proxy*—responds to requests on behalf of the CoAP IoT server. By using this module, security tasks like authentication/authorization and request rate-limiting can be delegated to another component.
- *Cross-proxy*—executes a protocol translation from the lightweight CoAP protocol to another protocol like HTTP. This proxy enables the protocol translation design pattern which is described in this chapter.

The forward and the reverse proxy can also cache the CoAP responses in order to deliver faster responses and to reduce the IoT network load. The security attack applicable to the cache feature is the cache poisoning attack, where an attacker injects a malicious CoAP response in a proxy and that response is further distributed in the IoT network until the cache is purged. A protection mechanism for the proxy cache poisoning attack is authenticating the CoAP server that delivers a response. Another protection mechanism is limiting the maximum cache lifetime according to the IoT network security policy, in order to satisfy a trade-off between security and network optimization. When designing a reverse proxy-based IoT architecture, this module can be used as DTLS terminator, handling all DTLS connections and offloading the servers from any security tasks. In this scenario, a reverse proxy can serve as a PEP, authenticating, and authorizing the clients and redirecting the requests to the configured servers based on the CoAP URI or using another policy. The forward proxy can serve as a PEP in controlling the incoming CoAP responses: it can rate-limit the responses according to the client IoT device capabilities or it can implement a filtering mechanism by blacklisting a suite of CoAP servers.

6.3.2.3 Multiple Central Brokers Pattern

An IoT network can be segmented in multiple trust domains, where a broker can host a certain number of IoT devices. In some cases, there is a requirement to interconnect different trust domains and apply a suite of security policies to the traffic flowing from one domain to another. Also, IoT networks can be geographically dispersed, thus facing the same scenario of interconnecting different trust domains. A design pattern that addresses this requirement is the multiple central brokers pattern, also known as *bridging pattern*. This design pattern is illustrated in Fig. 6.4. In a publish–subscribe network, the multiple central brokers pattern is translated into the following functional and security requirements:

- A publish message from one trust domain must have the possibility to flow in another trust domain.
- An IoT device must have the possibility to subscribe to a topic from another trust domain.
- An administrator must be able to configure the central broker bridging feature.
- An administrator must have the possibility to configure the security parameters for the publish–subscribe bridging: the bridging authentication and the bridging channel confidentiality.

In the MQTT protocol context, the bridging mechanics have a simple structure with one of the bridge sides acting as broker and as a client. Thus, a broker (bridge side) redirects the publish messages to the other bridge side. Also, the same broker subscribes, as a regular MQTT client, to the other bridge side for a suite of MQTT topics, then redirecting the received messages as generated by the local network. In terms of security, the bridge side acts as a regular MQTT client, thus the security paradigms from the central broker design pattern can be translated into this design pattern. Because the bridged MQTT broker is also a security element that can inspect all the messages, this module can also enforce ACL rules based on what traffic

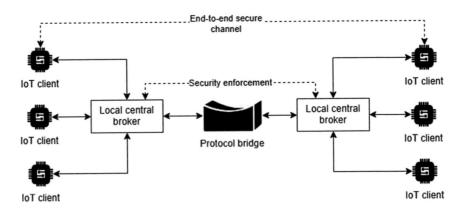

Fig. 6.4 Multiple central brokers design pattern

can be redirected from one side of the bridge to another. For instance, the MQTT *Mosquitto* solution allows defining bridge redirect ACLs based on MQTT topics. Thus, a security policy can control the topics that flow from one trust domain to another, while there is also the possibility of remapping the topic from one trust domain to another topic in the second trust domain. From the security point of view, the topic remapping hides the contextual information (e.g., the data topic might have a name that discloses private data from a trust domain) from the primary trust domain and transforms the information according to the authorization level of the second trust domain. An important feature of the MQTT protocol is the QoS capability, a feature that can be configured for bridging scenarios. Thus, a message distributed with a high QoS value in the local network can be transferred with a lower QoS value in another trust domain via the bridge.

6.3.2.4   Delegated Trust Pattern

In an IoT network, the central broker may have all the information required for executing security tasks like authentication and authorization: identity repositories, authentication tokens or ACLs. Another scenario is the one where the central broker delegates the actual authentication and authorization process to another entity and acts based on the received response. This security design pattern has the advantage of decoupling the decision point from the enforcement point, allowing the network to grow by adding several brokers querying the same security back-end. In a web context, the de facto standard for the delegated trust pattern is OAuth, but in the IoT world the UMA (User-Managed Access) technology gains continuous traction because it addresses the requirements of a constrained environment. Similar to the previous design pattern, the security operations in the delegated trust pattern can be executed at the application layer or at a lower communication protocol layer. The UMA technology is an application layer protocol because it uses the CoAP packets for transport, thus resembling the classical HTTP trust delegation method. A protocol-independent delegated trust pattern can be obtained by using the RADIUS authentication and authorization protocol. In this scenario, the central broker acts like a NAS (Network Access Server) and uses the IoT device identity to query a RADIUS server in order to obtain the authentication status, along with the authorization attributes. Moreover, the central broker can be an active authentication module in an 802.1X authentication scenario, where the broker relays the EAP packets between the IoT device and the RADIUS server. Even though 802.1X is not a popular protocol among the IoT devices, the central broker can apply an *authentication bypass* technique and extract the identity from a network packet, by applying a user-defined filter, and relay that identity to the RADIUS server, in order to obtain the authentication status. The RADIUS protocol has a simple structure, being transported using UDP packets and following a request-response model. Also, the protocol supports retransmissions, thus being adequate for lossy IoT networks. Another advantage of the RADIUS protocol is its ubiquity, this protocol being already extensively used in Wi-Fi networks for authentication/authorization scenarios. In the 802.1X context, an

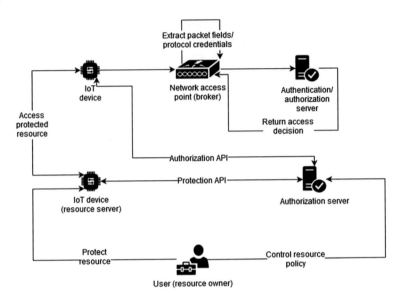

Fig. 6.5 Delegated trust design pattern

authentication bypass mechanism is applicable in the scenario of devices that are not 802.1X capable, the most known mechanism being the MAB (MAC Authentication Bypass). The MAB authentication uses the device MAC address as identity (802.1X being a Layer 2 protocol), but for the IoT context, a possible solution could be one in which a dedicated authentication agent runs on the broker side and extracts user-defined fields from network packets, the extracted data set representing the device identity. In such a scenario, the central broker does not play the role of a general NAS, being the network access point only for a certain service—like MQTT messaging. The user-defined filter can extract information like MAC address, IP address, and even MQTT credentials and wrap the data into RADIUS attributes. Figure 6.5 presents the user-defined filter authentication solution in the context of the delegated trust pattern. On the broker side, there is an agent parsing the network packet by opening a raw socket. This agent supports various *layers* and each *layer* can output an identity information that is encapsulated in a RADIUS attribute. The network packet information does not have a standard RADIUS attribute equivalent, therefore it is encapsulated by using a RADIUS VSA (Vendor-Specific Attribute). The authentication agent is a RADIUS client and after it receives the IoT device authentication status, it transmits this information to the network service (e.g., MQTT broker) by means of IPC (Inter-Process Communication) mechanism. The connection between the network parses agent and an MQTT broker can be realized using an authentication plug-in that acts only as an identity repository, storing a list of authenticated usernames. Because the network parser agent inspects the packet, in case of a TLS protected MQTT channel, application protocol information fields are not visible. In this situation, the agent can collect the decrypted protocol information from each network service (e.g., MQTT, CoAP or similar services) and apply the layer filter to the received packet. Being the de facto standard for the delegated trust pattern in the web world, OAuth is also a basis protocol for UMA. The main UMA concepts are the resource owner, the protected resource, and the entity accessing that protected resource. UMA allows the resource owner (can be a user or an IoT device) to control the protected resource access by multiple entities.

A classical scenario for UMA delegated trust pattern is the one of a Smart Home network, where devices managed by different manufacturers need to access the user private information.

The UMA OAuth profile has the following main phases [16]:

- *Protect a resource*—the protected resource is registered to the authorization server, with a list of security policies required for the resource access. The authorization server and the resource server are communicating through an OAuth-based API.
- *Get authorization*—the IoT device sends an access request to the resource server and obtains an authorization token by using an OAuth-based authorization API.
- *Access a resource*—the IoT device delivers to the resource server an authorization token that allows the data access.

In the UMA security framework, the resource and the authorization server are using a TLS protected HTTP API, which has the next phases:

- The authorization server issues credentials to the resource server.
- The resource server gets a PAT (Protection Access Token) from the authorization server.
- The resource server registers the protected entity to the authorization server.

A notable feature of the UMA OAuth profile is the flexibility in terms of resource protection. The owner can enforce this protection or the resource server can decide to register an entity for protection. The service provider must validate the RPT (Requesting Party Token) authorization level for each protected resource. The UMA authorization process consists of the following steps [16]:

- The IoT device tries to access a protected resource.
- If the IoT device does not transmit an RPT, the resource service provider requests a token from the authorization server and relays it to the IoT device.
- The IoT device uses the previously obtained token to obtain access to the protected resource.
- If the IoT device authorization is successful, then the authorization server generates an RPT for the desired resource.
- The IoT device uses the previously obtained RPT and sends it to the resource server in order to be granted access to the protected resource.

If the UMA RPT is self-contained, then the resource server can take the access/deny decision based on the token value, otherwise, the resource server must query the authorization server in order to take the grant decision. A particular UMA feature is the interactive mechanism, where a user or an IoT device can be contacted in near real-time to participate in the authorization process. UMA addresses a broad range of IoT architectures, proposing in [16] various extensibility profiles. UMA specifications propose HTTP as the transport protocol but also mention CoAP in order to address the resource-constrained IoT deployments. The main advantages of UMA in the IoT context, as presented in [17] are the following:

- The user (owner) interaction is not required for each authorization process.
- The user (owner) can query the IoT device authorization status.
- The user (owner) can dynamically modify the access parameters.
- There is a central entity handling the authorization process from various trust domains, thus offloading this task from the trust domain central broker.

An IoT scenario showing the UMA role separation is the one of door-lock authorization system [17] that has the following stakeholders: the door-lock is the protected resource, the door-lock purchaser is the resource owner and the authorization server can be the central broker. A series of security considerations regarding the OAuth and UMA in the IoT context are the following:

- DTLS must be used to validate the server components and to mitigate MiTM attacks by ensuring the transport channel confidentiality.
- Short-lived authorization tokens mitigate the risk of granting access to an IoT device that can change its authorization status. Moreover, short-lived tokens are suitable for security architectures that lack (real-time) revocation mechanisms.
- Scoped tokens granting access only to a well-defined set of resources provide a fine-grained access control solution.

6.3.2.5 Network Protocol Aggregator Pattern

The IoT networks are populated with low-end sensors having the unique task of transmitting the acquired data to an upstream module. A common mean of delivering the sensed data is to have a network node, called *aggregator*, that acquires data from multiple end-node sensors, aggregates and process it, then delivers the information to the upstream network element. This sensed data delivery scenario can be handled by means of the network protocol aggregator pattern, which specifies common design elements for the IoT data aggregation scenario. From the security perspective, the IoT aggregator is an active network element that can modify data, thus it has to be trusted by both the endpoint sensors and by the upstream module. General architecture of the aggregator design pattern is illustrated in Fig. 6.6. If compromised, the aggregator can tamper the sensed data and, thus it can easily execute false data injection attacks. Also, a compromised IoT aggregator can disrupt the network services by executing a black-hole attack on the sensors' transmitted data.

From the functional point of view, the IoT aggregator usage reduces the number of transmitted network messages, thus reducing the network congestions and battery drain scenarios. Another aspect of using the aggregator pattern is filtering and normalizing the sensor data:

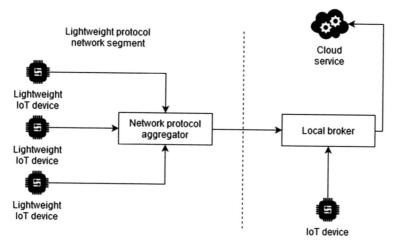

Fig. 6.6 Architecture of aggregator design pattern

- The IoT aggregator can compute an average of the received sensor values.
- The IoT aggregator can change the sensor data format in order for the information to be consumed by the upstream modules.
- The IoT aggregator can filter the sensor data in order to reject invalid values and report the faulty sensors.

By analyzing the IoT aggregator concept in the context of the most employed IoT communication protocols, we can identify this design pattern as built in the MQTT-SN protocol. For the CoAP protocol, the aggregator design pattern can be implemented using a CoAP forward proxy module that intercepts the sensor requests and processes them before delivering the information to the CoAP server.

MQTT-SN is a UDP-based lightweight variant of the TCP-based MQTT protocol, employed especially in IoT networks populated with resource-constrained sensors. In the MQTT-SN paradigm, the protocol aggregator design pattern is implemented using a gateway element linking the MQTT-SN sensors to the MQTT broker. The MQTT-SN gateways can be of two types:

- *Aggregating gateways*—this type of gateway collects multiple MQTT-SN messages and sends it to the MQTT broker one MQTT message (this type of gateway implements the aggregator design pattern).
- *Transparent gateways*—this type of gateway is a simplified version of an aggregating gateway and has the sole task of translating a message from MQTT-SN to MQTT (from UDP to TCP).

The MQTT-SN protocol is a simplified version of MQTT, having a series of message types associated with the gateway module:

- ADVERTISE—this type of message is periodically broadcasted by the gateway in order to advertise its existence in the network.

- SEARCHGW—this type of message is broadcasted by the sensor (client) in order to search for a gateway.
- GWINFO—this type of message is broadcasted as response for a SEARCHGW message.

Because the gateway discovery mechanism is handled using broadcast messages, the radius of the message is indicated by the client to the transport layer when transmitting SEARCHGW messages. Being a lightweight protocol, MQTT-SN specifications do not mention any in-protocol security features, delegating the security tasks either to the lower layer transport protocol (e.g., by means of DTLS) or to the upper layer application protocol (payload), that can secure the messages in an application-specific manner. For securing an IoT aggregator architecture, authentication and integrity layers must be implemented on top of the sensor-aggregator gateway channel. The connection between the aggregator and the broker can be secured using the paradigms from the central broker design pattern, the communication session being a simple client-broker link, in the publish–subscribe network scenario. Moreover, the central broker must authenticate the aggregator, because this module modifies the data received from sensors and it is delegated with the security task of authenticating the sensors and verifying that the received data does not tamper. The security layer between the aggregator and the endpoint sensors may differ, depending on the IoT application use-case and the sensors computing capabilities:

- *No security*—the endpoint sensors do not have the capability of implementing any security mechanisms. In this scenario, the aggregator and the sensors must be in the same trust domain or the aggregator must filter the untrusted data by using a consensus mechanism (assuming that an attacker cannot compromise the majority of the sensors in order to change the average aggregated data). Also, the sensors are offloaded from the task of securing the messages, but the aggregator must implement additional logic in order to achieve the required security level.
- *Message*—the IoT application requirements might not need sensor data confidentiality, taking into consideration the fact that the data may be publicly available. However, data integrity must be assured, in order to mitigate false data injection attacks and to mitigate any transit data modifications. In the MQTT-SN context, data integrity can be implemented at the application level, by using JWT signatures. This security scenario requires a trust relationship between the sensor and the aggregator, either by using a pre-shared key or an asymmetric key pair.
- *Message integrity and confidentiality*—if the sensor acquired data is not distributed in the IoT network, the communication channel between the sensors and the aggregator must be confidential. The transport channel confidentiality can be assured by using a DTLS session as specified in the central broker design pattern, or the sensor data payload can be encrypted using a JWT. This security scenario requires a trust relationship between the sensor and the aggregator. Depending on the required security level, the sensor and the aggregator might use the trust key to generate a session key, that encrypts the communication channel.

The MQTT-SN security protocols must take into consideration the limited MTU of the underlying transport protocol, digital certificate-based solutions not being recommended due to packet fragmentation.

6.3.2.6 Network Protocol Translator Pattern

An IoT network employs lightweight network protocols at all the communication layers, starting with 6LoWPAN and ending with MQTT and CoAP. Also, an IoT network is fragmented in terms of communication protocols, with proprietary devices with different computing capabilities. The goal of an IoT network is to deliver the locally acquired data to a Cloud component that can manage the use of data and send actuation commands to the IoT network. All these IoT network characteristics require network protocol translation, from a lightweight protocol to a complex protocol and vice versa. This requirement is implemented with the aid of the network protocol translator pattern, which gives directions for handling the main issues of this requirement, including the security aspects. A relevant example for the protocol translation pattern is the previously described aggregator pattern, where the MQTT-SN gateway translates messages from the lightweight MQTT-SN to MQTT, in order for the messages to reach a central broker. The protocol translator pattern has the primary purpose of moving the application protocol payload from one transport type to another. From the security point of view, this design pattern feature brings many advantages because the protocol translator module does not have to modify the application protocol payload, thus keeping valid any end-to-end security channel. This feature does not necessarily mean that an IoT protocol translator is a passive network module that is transparent from the security perspective. The protocol translator operation can be a complex process and a compromised protocol translator might change sensitive packets fields from the newly converted application protocol. For instance, in the case of MQTT-SN and MQTT, a protocol translator attacker can change the QoS value when converting the protocols, thus executing a protocol field attack and propagating false information to the upstream network segment. By maliciously modifying the packet header fields, an attacker can control the number of retransmissions and impact directly the battery consumption. In Fig. 6.7 is presented the network protocol translator pattern. In the CoAP context, the protocol translation process is executed by means of a cross-proxy, the most encountered translation scenario being the one from CoAP to HTTP.

Being a simplified version of HTTP, the CoAP to HTTP translation is not a complex process, this being a CoAP design feature with CoAP headers having an equivalent HTTP header within a limited subset. The protocol translation pattern allows IoT network management and visualization by means of a web interface by converting CoAP to HTTP messages. As mentioned in [14], the CoAP proxy can have the following types:

- *CoAP to HTTP proxy*—allows an IoT device to execute HTTP requests. The CoAP request must contain the Proxy-Uri or Proxy-Scheme option with an HTTP value.

Fig. 6.7 Architecture of
network protocol translator
design pattern

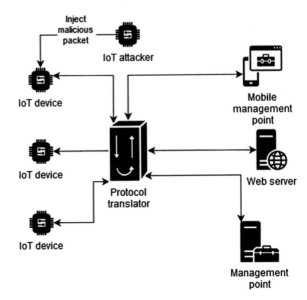

- *HTTP to CoAP proxy*—allows a Cloud module to execute CoAP requests. The HTTP request must contain a Request-Line option with the CoAP value.

The CoAP translation is applied only to the request/response model of CoAP, which means the common subset of these protocols. The CoAP to HTTP translation uses the following HTTP methods [14]:

- GET—the IoT device can obtain a representation of an HTTP resource and the translator must set the Content-Format according to the HTTP resource type. The client can include an Accept field in the CoAP request in order to indicate to the translator module the preferred Content-Format. This protocol translation operation includes the ETag cache control mechanism, resembling the HTTP ETag paradigm.
- PUT—allows an IoT device to create or update an HTTP resource. The translator returns a 2.01 status when the resource is created and a 2.04 status when the resource is updated.
- DELETE—allows an IoT device to delete an HTTP resource. The translator returns a 2.02 status when the resource is deleted.
- POST—allows an IoT device to manipulate an HTTP resource by sending it to the HTTP server a payload to be processed.

The HTTP to CoAP translation uses the following CoAP methods [14]:

- GET—the Cloud component can obtain the representation of a CoAP resource. The response must contain a max-age parameter which indicates the response freshness. The response can also contain an ETag header value.

- HEAD—even though CoAP does not support the HTTP HEAD method, the translator can simply return the HTTP response without a payload.
- POST—allows the Cloud component to send a representation enclosed in the request body to the CoAP server. The CoAP server processes the request and can return an appropriate response based on the resource creation status.
- PUT—allows the Cloud module to update or create a CoAP resource.
- DELETE—allows the Cloud module to delete a CoAP resource.
- CONNECT—this method is not supported on the translator side, due to lack of support for TLS to DTLS tunneling.

The protocol translation pattern must handle a suite of security attacks, among which is the node manipulation attack. This type of attack is a form of cross-protocol attack, where a legitimate network node processes the packets sent by a manipulated IoT node or vice versa. This attack uses the IoT device as a relay module, with the attacker sending a manipulated packet to another victim node. Another variant of this attack consists in the scenario where the IoT device processes a packet originated from another protocol. A classical CoAP cross-protocol attack [14] is the CoAP-DNS attack, where the attacker executes a spoofing on the source address and a DNS server sends an unsolicited response that is interpreted by an IoT device as a CoAP packet. The CoAP-DNS attack allows the attacker to bypass a firewall rule, normally blocking the communication channel from the attacker to the victim. This attack impacts the IoT device battery life, the embedded module being forced to process and parse a manipulated CoAP packet. A mitigation technique for the cross-protocol attack is the correct packet parsing if the protocols are different enough. Another mitigation technique is an anti-spoofing mechanism, this being the trigger of the CoAP-DNS attack. Being based on UDP, CoAP is vulnerable to various attacks, due to a lack of context of the UDP protocol. By using a DTLS session, a CoAP node can establish a session with the requestor, while authenticating the peer node. As mentioned in [14] a DTLS session can mitigate a cross-protocol attack only if hosts a single type of protocol.

6.3.2.7 Application Protocol Segmentation Pattern

An IoT network can be populated with both trusted and untrusted devices that can consume different network services. The interaction between these IoT devices must be designed using an isolation model, where untrusted devices cannot reach the trusted devices. Taking into consideration the complexity of an IoT network, the interaction between devices can be modeled as a suite of layers, where each layer represents an application protocol. Thus, in order to design a granular security solution, the segmentation mechanism must be implemented for each protocol (layer). By using the protocol segmentation design pattern, the IoT network design gains flexibility in terms of device interaction. For instance, two devices can have visibility when using a client-server protocol like CoAP but are isolated when using a publish–subscribe protocol like MQTT. In a classical network, the device isolation is achieved by means of the well-known VLAN (Virtual LAN) concept, which allows overlaying several virtual networks on top of the same physical infrastructure. The VLAN paradigm works by adding a VLAN tag (2 bytes identifier) on the Layer 2 segment of the network packet. The VLAN mechanism is not considered feasible for the IoT networks because the implementation costs, in terms of packet overhead and hardware complexity, is too high. Added to these characteristics is the static nature of the VLAN concept that does not address the dynamic character of an IoT network. Considering this, a possible solution may be the protocol segmentation design pattern which resembles the VLAN concept. The protocol segmentation pattern can be applied specially to publish–subscribe protocols, having the main purpose of blocking unauthorized publishers in injecting false data into subscriber devices. The segmentation process must be implemented using a filtering mechanism that classifies the unauthorized/malicious publishers and blocks the traffic from a publisher to a subscriber. This filtering mechanism must be instantiated for a publish–subscribe session, being aware of communication context information, like TLS certificates used in authenticating the client to the broker. The main advantage of the protocol segmentation pattern is the possibility to have a hybrid IoT network, where untrusted devices can dynamically join the network without affecting the security of sensitive devices. For the protocol segmentation pattern, the MQTT vTopic (virtual topic) concept may be used, as a security paradigm defined in [18]. The MQTT vTopic acts like a normal MQTT topic, hiding the filtering security mechanism for both the clients (publishers, subscribers) and the core MQTT broker. The process of creating an MQTT vTopic consists of the filtering policy installed by an authenticated subscriber on the broker side: the IoT device sends to the broker a list of identifiers or a set of attributes that must be owned by the publisher device, in order for the message to reach the authenticated subscriber. Taking into consideration these characteristics of the MQTT vTopic, we can conclude that the protocol segmentation pattern is based on the central broker pattern, the broker acting as a PEP and enforcing a filtering mechanism. The device publishing a message to a vTopic, executes a normal MQTT publish process without having to perform any security tasks, the vTopic structure being represented by the tuple (MQTT topic, filtering policy). Considering the vTopic structure, multiple IoT devices can subscribe to the same MQTT topic, but only a subset of them can transform that topic into a vTopic. The implementation of the vTopic structure follows a modular approach with the security functions being separated from the core broker logic by using a security plug-in. The core broker communicates with the security plug-in by using a callback mechanism. Mosquitto, an open-source MQTT broker offers a security plug-in that consists of a Linux shared object library. Such a solution has a well-defined interface that can be easily extended with a set of functions in order to implement the vTopic mechanism. By separating the vTopic implementation along with other classic MQTT security solutions like ACLs, the solution maintenance is eased, allowing even porting the security module to another MQTT broker solution. The vTopic architecture is presented in Fig. 6.8.

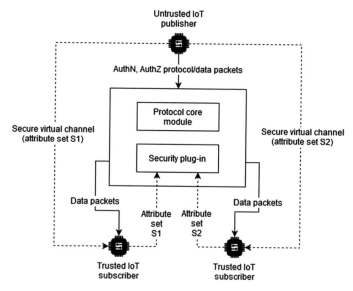

Fig. 6.8 Architecture of the MQTT vTopic concept

The security plug-in has a back-end source (Redis database) for provisioning the filter rules. Each MQTT publish message is transmitted from the core broker to the security plug-in before being redirected to the subscribers. If the security plug-in detects a vTopic, then it applies the filtering policies and returns the status to the core broker, which redirects the message only if the filtering was successful. If the vTopic filtering fails, then the core broker drops the packet and logs the security event. As mentioned in [18], the vTopic plug-in callback can have the following form:

typedef int (*FUNC_auth_plugin_vtopic_filter) (void *user_data, const char *source_id, const char *destination_id, const char *topic);

6.3.2.8   QoS Pattern

IoT networks are populated with a wide range of embedded devices, each having different capabilities and purposes. Publish–subscribe paradigm, the leading IoT network concept, employs a central broker for delivering messages from publishers to subscribers. The broker can be a limited resource device that must handle messages from an entire IoT network. Taking this into consideration, the broker can handle a limited number of messages in a given time interval, thus leading to a QoS-based network architecture. Particularly, the MQTT protocol has a built-in QoS mechanism with 3 values:

- *QoS 0*—the MQTT packet is delivered at most once.
- *QoS 1*—the MQTT packet is delivered at least once.
- *QoS 2*—the MQTT packet is delivered at exactly one.

As previously explained, a higher QoS value enables a reliable MQTT transport channel, with the price of an overhead, in terms of additional exchanged messages between the client and the broker. The client–server equivalent protocol (CoAP) does not provide any QoS parameters for differentiating the network packets, so any QoS logic must be implemented at the application layer or at the network layer (e.g., IP differentiated services). While the QoS feature provides better message convergence under high network load, it raises a series of security issues:

- MQTT does not provide any QoS-related authentication, thus the QoS value is under the client control.
- A client attacker can send messages with a high QoS value, thus executing a DoS attack on the broker side and blocking legitimate MQTT messages in being distributed in the required time interval.
- A client attacker can send messages with a high QoS value in order to drain the battery on the broker side and on the subscriber client-side, thus forcing these devices to exchange an additional number of messages.

The majority of IoT networks run real-time or near real-time applications, thus the broker must deliver the messages in a low latency manner. This requirement is highly affected by the QoS-based attacks because a malicious device does not have to exhaust all the broker resources in order to conduct a successful attack. In order to affect the delivery time requirement, the malicious device has to consume enough CPU cycles in order to delay the message transmission above a certain time threshold. The QoS design pattern offers a solution to QoS security issues in order to take full advantage of this MQTT built-in feature. A relevant use-case for the QoS pattern is the same scenario as the one from the protocol segmentation pattern: an IoT network composed of a group of both trusted and untrusted publishers that communicate with a group of trusted subscribers. In a manner similar to the protocol segmentation pattern, the group of untrusted publishers might inject high QoS messages into the network in order to disrupt the communication between the trusted devices and to increase the broker message delivery time. A possible solution for the QoS design pattern may be the QoS-based delay control mechanism, defined in [18]. The test scenario, described in [18], consists of an increasing number of publishers that are sending MQTT messages with a QoS value of 1 and 2 (the QoS 0 value was omitted because the transmission is not reliable). The message delivery time was measured as a timestamp difference between the publisher and the subscriber. As expected, a high number of QoS 2 messages impacts drastically the message delivery time, this being a mechanism that can be easily exploited by a group of attackers. The QoS pattern implementation splits the MQTT communication channel into two planes: a data plane, transporting the actual messages, and a security control plane, transporting the authentication and the delay report messages. At the core of the QoS pattern implementation stands a delay report solution from the trusted subscribers, which can report to the broker any unusual delays breaking the near real-time constraint of the IoT application. The broker starts dropping MQTT publish messages from untrusted devices until the network delivery time converges to the required time value. For the security control plane, a TLS channel with mutual authentication is used, in order to identify the trusted publisher and subscriber devices. For the packet dropping mechanism, the core broker transmits each received message to the security plug-in in order to obtain the relay/reject status. This message trap mechanism is implemented with the following extension to the standard Mosquitto security plug-in:

typedef int (*FUNC_auth_plugin_msg_trap) (void *user_data, const char *client_id, const char *topic, uint8_t qos, uint32_t payload_len, const char *payload);

The security plug-in consumes both the data plane and the security control plane messages: the QoS of the data plane messages is analyzed and the packets are dropped if required, and the security control plane messages are identified based on their topic value. Thus, if an authenticated device sends a message to a pre-defined control plane topic, the security plug-in decodes the query and registers the message delivery delay value.

An important security aspect of the QoS design pattern consists of spoofing attacks: IoT devices considered to be trusted can impersonate other devices, in order to execute malicious actions. The spoofing attack type spans across the communication stack, a main concern in the IoT context, being the application protocol spoofing. The MQTT protocol has a unique identifier in the broker namespace: the *client id* field that can be either auto-generated or distributed by using an MQTT out-of-band mechanism. A simple MQTT spoofing attack consists of changing the client id parameter in order to escalate authorization privileges on the broker side. Also, by forging an MQTT identity, a malicious device can execute a reputation downgrade attack by sending packets to unauthorized topics, thus causing the broker to log unauthorized actions for the victim MQTT client. In order to mitigate the MQTT spoofing attack, a possible solution is also presented in [18]: distribute unique MQTT client id to devices and embed the client id into the client certificate (this field can be set as a certificate Distinguished Name component, like serialNumber). Thus, if MQTT clients are authenticated to the broker using a TLS connection with digital certificates, the broker can verify the actual client id from the MQTT header against the client id from the certificate used in securing the transport protocol. By using this mechanism, the client id is handled by a trusted third party (which is the Certificate Authority) and the broker can rapidly detect this type of MQTT spoofing attack by employing a low overhead validation method.

6.3.2.9 Asynchronous Authorization Design Pattern

As stressed in the previously mentioned design patterns, the IoT device authentication and authorization is a critical aspect that must be handled in almost any IoT security scenario. In a dynamic environment like an IoT network, devices should be authorized based on a set of attributes. In a communication scenario intermediated by a third module, such as a CoAP proxy or an MQTT broker, the message sender/receiver can be authorized in an asynchronous manner, only when required. By using such a mechanism, an untrusted publisher can be challenged by a broker to prove a series of attributes when sending a message to a subscriber that accepts data only from authenticated devices. This design pattern brings two important advantages: the IoT network load is reduced, by exchanging authorization packets only when required, and the IoT device discloses its attributes only on a need-to-know basis, thus leading to an improved privacy solution. The asynchronous authorization pattern consists of the following sub-processes:

- An authenticated subscriber installs on the broker side a list of attributes.
- When sending a message to the previously mentioned subscriber, the broker initiates an authorization protocol in order to find if the publisher holds the required attributes.
- The publisher uses the required attributes in order to authorize the broker.

A flexible authorization mechanism can be built on the ABE (Attribute-Based Encryption) paradigm, as presented in [18]. In the ABE context, an external authority authenticates the device and issues a list of cryptographic attributes. By using the CP-ABE (Ciphertext Policy ABE) concept, a device may decrypt a message only if it is issued with the required attributes used to encrypt the message. For the authorization protocol, the broker generates a nonce and encrypts it with the attributes indicated in the subscribers' security policy. The encrypted message is then sent to the challenged device by using an MQTT publish message. After receiving the message, the device uses the issued attributes to decrypt the nonce and sends back to the broker, via another publish message, the plaintext value. This CP-ABE-based challenge–response protocol also employs the MQTT security control plane presented in the QoS design pattern section: the challenged device must use a pre-defined MQTT topic in order to exchange the security challenge messages with the broker. An implicit advantage of this authorization mechanism is the possibility to execute a group authorization, by only sending a single challenge from the broker side. The challenge publish message will reach every subscriber, that will respond back with the decrypted information (it is assumed that once a device decrypts the authorization token, it will not pass it to another device that does not hold the attributes). Another advantage is that this mechanism does not require an online attribute authority participating in the authorization process. The attribute authority must be online only when identifying and issuing the attributes and the secret key to the device.

6.3.2.10 Traffic Analytics/Packet Filtering Pattern

IoT gateways are core IoT network modules, being the intermediary element between the endpoints and the Edge network. Also, gateways do not have the same resource constraints as the endpoint IoT modules. Being capable of hosting various application protocols (client–server, publish–subscribe etc.), the gateway is the central point where traffic analysis can be executed or traffic filtering policies can be applied. As stressed in [19], the gateway/edge modules are a feasible solution for offloading cloud processes in order to benefit from better data locality (this is also the case of real-time IoT traffic analysis). In a publish–subscribe network, the gateway can relay messages between the local network devices and can also relay messages between the local devices and the upstream Cloud module. Network packet analytics help in detecting attacks from authenticated devices, attacks that cannot be detected by using cryptographic mechanisms. Also, by implementing a traffic analysis design pattern, an administrator can have visibility over the IoT nodes behavior and can rapidly detect potentially malicious traffic that does not follow a required pattern. The traffic analytics/filtering pattern can also be implemented on the client–server proxy modules, where each packet can be analyzed in order to detect malicious behavior. A packet filtering or analysis policy can be also based on metadata, like the number of packets per second or the time of the day when the packets are sent. Regarding the software structure implementation of this design pattern, it can be split into two analysis/filtering methods: a userspace module which is integrated with the application protocol process (e.g., MQTT broker, like Mosquitto) and a kernel mechanism which allows rapid packet analysis before the packet reaches the userspace. From the efficiency perspective, the kernel filtering method is superior because a packet classified as malicious can be rejected in an early stage of the Linux packet processing pipeline (before the skb structure kernel structure is allocated). Thus, the kernel packet processing method avoids memory allocation and application protocol process scheduling along with additional packet parsing on the process side. The main drawback of the kernel filtering method is that it cannot apply pattern matching algorithms on TLS/DTLS encrypted data, the encryption keys being stored on the process side. An alternative to the kernel filtering method is analyzing the packets on the userspace side by using a module attached to the application protocol process: in this manner the traffic is decrypted by the process and transmitted for analysis to another module. The userspace analysis/filtering method can be implemented by means of a Mosquitto plug-in, as described in the QoS design pattern section, by using a message trap function that allows the transmission of a packet from the broker side to the plug-in side. For the kernel filtering method, an in-kernel virtual machine that runs eBPF instructions (also used by the well-known tcpdump solution) may be used. An eBPF structure permits a limited set of instructions, that are statically analyzed before being injected in the kernel. eBPF instruction set enables the network packet parsing along with the possibility to design an early packet drop mechanism if the packet is classified as illegal. The XDP/Iovisor open-source project permits injecting eBPF instructions using a Python front-end. Thus, a packet filtering program, written in the C programming language, can be compiled into eBPF instructions using the *clang* compiler and installed into the kernel by using a Python process. A particular scenario, where the traffic analytics pattern can be useful, is the DoS attack caused by multiple MQTT PUBLISH packets or by CoAP multicast responses. In these cases, the traffic analysis pattern can rapidly detect the illegal network behavior and can begin dropping packets, in order to protect the resource-constrained devices. In Fig. 6.9, the traffic analysis/filtering design pattern is presented.

As it can be observed, the system is managed by a controller that can install packet matching rules in both kernel and userspace. Thus, the packet filtering modules act as agents and statistics collectors, transmitting the acquired data to the controller that can take complex decisions, in order to correct the network behavior. The statistics collection can be easily achieved by both the kernel-based eBPF solution and the userspace-based Mosquitto plug-in. Taking into consideration that this design pattern

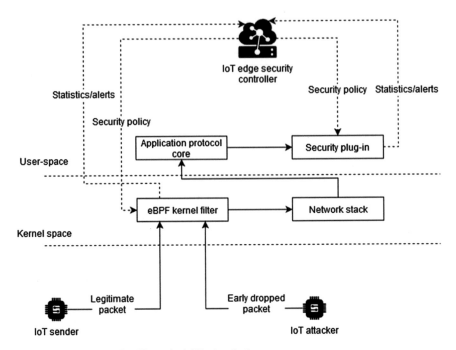

Fig. 6.9 Architecture of traffic analysis/filtering design pattern

is implemented on the gateway side, the communication between the controller and the gateway agents can be realized using a TLS protected REST API, with the gateway exposing endpoints for configuring the analysis/filter matching rules and controller exposing REST API hooks for asynchronous events notification from the gateway side.

6.4 IoT Cloud Platform Design Patterns

Together with the rising popularity of Cloud-based services, manufacturers of IoT solutions have migrated their online management and integration solutions to a Cloud perspective, bringing no interference in the manner services are provided to users, but improving the scalability, accessibility and availability of these services. Yet, as in the case of any other integration of two or more concepts, Cloud platforms for IoT systems have a certain amount of deficiencies, mostly caused by security or interoperability aspects. Some of these deficiencies can be addressed through an accentuated need for the better design of the final solution that will be deployed, by focusing on design and development patterns adequate for a specific type of IoT system. Depending on these types, there are three main profiles that a Cloud platform can have:

- *Consumer platform*—oriented toward regular users, thus leading to an integration of several types of devices with different characteristics that need to be taken into account. Collected data needs to be processed and stored individually for each user and access to services or resulting information needs to be monitored and regulated based on users' preferences.
- *Industrial platform*—focused on the central management of multiple devices located in a factory, a farm, a power plant or other compound that runs an industrial process. Even though devices can be systematized in classes of functionalities that contain devices with similar features, they can also be divided into new or legacy devices, thus hindering their integration process. Data can be collected, processed, and stored, while permissions are kept to a minimum, information being circulated only inside the compound.
- *Enterprise platform*—can be seen as a particularity of an *Industrial platform*, but designed for offices or organization buildings. Data is collected with the intent of optimizing resource consumption (e.g., electricity, water) and automating recurring tasks (e.g., temperature control). In this case, permissions to data are simple, the building administrators being the only persons that can have access to it.

Identifying the desired profile of a Cloud platform is the first step in constructing it and, by implementing specific design patterns, many possible deficiencies could be eliminated directly from the design phase. Even though there are many challenges when designing a Cloud platform, such as availability, resilience or data management, security can be considered a primary requirement in the design process. Given the amount and importance of private data that exists in IoT systems, Cloud platforms for IoT will not be overlooked by attackers, therefore adapting the security requirements to the needs of users and, implicitly, to the platform services, is a must. In the following part of this section, several design patterns for Cloud platforms will be analyzed, more from a security perspective, rather than purely from a functional one.

6.4.1 Security Division Pattern

When migrating their central management platform in Cloud, IoT systems vendors have three options to choose from: a *Public Cloud* (resources are owned by a third party and delivered over the Internet), a *Private Cloud* (resources are directly owned and operated, without the involvement of a third party) or a *Hybrid Cloud* (a combination of a *Public* and *Private* Cloud, where services can be divided depending on the hardware requirement or volume and sensitivity of data). A particular design pattern for the Hybrid Cloud profile, from a security perspective, can be a separation between the application-specific security requirements and the security of general data. This separation can be defined by decoupling the major security requirements of the IoT system through a security-based platform on the Private Cloud that will enforce data filtering, access control, and device/user authentication and authorization. In this manner, the Public Cloud will act mainly as a storage provider and, in

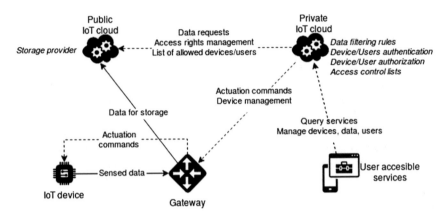

Fig. 6.10 Architecture of the security division design pattern

case of emergency, as a duplicate of several functionalities defined on the Private Cloud. The security-based platform on the Private Cloud becomes a gateway between devices or user requests and the storage module, as presented in Fig. 6.10.

This design can be derived and extended to allow integration with a fully functional Cloud IoT platform, the software-based platform being transformed into a security service, still acting as a gateway between devices/user and Cloud services. In this scenario, the roles of the security service could be reduced to specific requirements, such as data filtering, the remaining ones being ensured by the Cloud IoT platform, or could remain unchanged, defined as complementary to the ones the Cloud IoT platform provides. By using the design in this extended approach, it could be adapted not only for a Hybrid Cloud profile but for any type of Cloud profile. The security service becomes completely decoupled from the Cloud paradigm, therefore it can be implemented either as a web-gateway for user devices to connect through or it can be integrated into a local or Edge gateway, specially designed to support connections from the base layer of devices/sensors. This decentralization of the IoT system management framework can provide a certain degree of adaptability, making it easy to be integrated into either one of the three Cloud platform types (Consumer, Industrial or Enterprise), depending on the needs that have to be met.

6.4.2 Digital Twin Pattern

Services exposed to users by an IoT system need to have uninterrupted access not only to data sent by IoT devices but also to the status readings for these nodes. Taking into account that there are multiple design patterns about how connections with IoT devices can be made and many of them tackle the constraint of battery life, situations may exist when Cloud services will not be able to quickly analyze the status of a node. To address this issue, many Cloud providers have considered

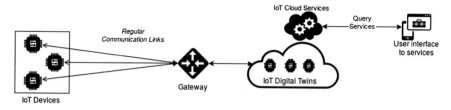

Fig. 6.11 Architecture of the digital twin design pattern

the implementation of a virtual representation of an IoT device (Digital Twin design pattern [20]), adapted to the characteristics of the real node. As presented in Fig. 6.11, this layer of virtualization can act as a cache for commands or data requests being initiated by users, by temporarily storing the last bulk of data sent by a node. Also, by cloning characteristics of the real device, a virtual node can offer similar user-experience, such as giving users access to specific ports or hardware configuration parameters when creating queries. Another functional aspect that this design brings is the possibility of creating a virtual design of an IoT system, without having any devices physically connected to it, for the scope of defining communication flows or implementing security mechanisms. After the successful deployment of such a virtual architecture, its elements will be filled with real data after each physical device will be connected to the Cloud platform and an association between these virtual and physical elements will be made. From a security perspective, this design pattern drastically reduces the risk of outputting erroneous data to users, given that data stored in the virtual node is not new data, but a previously verified one. Combining this pattern with the Health Endpoint Monitoring design pattern, described in the following paragraphs, users are enabled with valid and updated data.

The widespread of IoT devices in different layers of society brought an increase in efficiency in daily tasks, but because different IoT systems are not always interconnected, an IoT system functionality is limited to a given feature set. Thus, without the capability of integrating data from outer sources and use it to improve the accuracy of provided services or auto-regulate, an IoT system can become difficult to be managed in certain circumstances. The Sensing-as-a-service (S2aaS) paradigm allows an IoT system to aggregate complex data and to present them to users, even though it does not physically have the required hardware module for collecting that type of data. Among the various advantages that this paradigm brings, there are also some economic ones. The main one is the reducing of costs when an IoT operator wants to update his provided services. Instead of upgrading all his IoT devices, he can simply install new ones that offer the method of collecting new data. By being compatible with the S2aaS paradigm, these new devices will be able to integrate into the existent communication network and aggregate data collected by existing devices. If different IoT domains can establish a trust relationship and ensure the security of data and communications, then a large-scale implementation of services can be implemented.

6.4.3 Secure Design Through Microservices

When combining the adaptability and security aspects of a Cloud platform, its modularity feature becomes a determinant criterion in the design and implementation cycles of the platform. At this point, the concept of *microservices* becomes a targeted architecture to build a Cloud platform upon. The microservices architecture implies the creation of a set of loosely coupled services, each one with a specific function implementation, collaborating through the use of either synchronous protocols, such as HTTP/REST, or asynchronous protocols, such as AMQP. This way, by standardizing the method of inter-service communication, services can be developed independently and deployed in different layers of the IoT system (taking as example the design pattern mentioned above, the suite of security microservices can be split dynamically between the Hybrid and Public Clouds). Although this separation of functionalities brings many advantages, it also has some limitations regarding the method of communication with various services that expose different entry points. A possible solution to this issue can be the integration of an API Gateway pattern. It involves creating a single entry point that will handle requests coming from clients or their devices, either by simply routing them to adequate services or by parsing the requests and identifying the group of services as the intended destination of the request. This software element can also be modularized, by exposing APIs adapted for different client types, thus limiting the attack surface in case of a rogue device. From a security perspective, an API Gateway can be decorated with several security mechanisms, such as checking that a device or a user has the required authorization to make a certain request.

Microservices are becoming tightly connected to application development for IoT environments by increasing the level of control and abstraction of the primary functionalities required by such a system. Yet, a too general ramification of microservices can prove to be a rigid structure in the overall architecture, while a too deep ramification can be hard to manage. Addressing the functional division of microservices, the following classes of services can be considered sufficient as the core of a Cloud IoT platform: Security, Data Management and Resource Monitoring. This separation does not include other fundamental services, like the ones that provide means of communications, authentication, and authorization of devices/users or storage management. In [21], Microsoft proposes a set of design patterns for a balanced implementation of a Cloud platform, patterns that are implemented also in Azure, the Cloud platform of Microsoft. A part of these design patterns is also reiterated in [22]. In the following paragraphs a part of these design patterns is adapted for usage in an IoT system and grouped under some of the functional classes previously defined.

6.4.3.1   Security Class

Federated Identity

The steady pace of IoT technology acceptance by the society and the continuous expansion of these IoT systems in the surrounding environment, led to an increase in the number of IoT providers offering a mixture of devices and sensors with the aim of improving the quality of everyday life. Yet, the lack of a rigorous standardization meant that any development or upgrade cycle for these devices was done under the supervision of the parent company, without the need of assuring a certain degree of interoperability with similar devices from other vendors. Even though a mixt Cloud platform can accept data from various IoT sources (different data format types from different vendors), surpassing the interoperability issues, trust is still a problem that needs a well-defined standard procedure. The Federated Identity pattern proposes the trust establishing not directly between the mixt Cloud platform and the heterogeneous devices, but between the mixt Cloud platform and the corresponding Cloud platform where each device is authenticated [23]. Therefore, the need to tediously translating a mixture of security features and parameters in a common set can be partially resolved [24].

Gatekeeper

In an IoT system, a Cloud platform will continuously receive data either from devices/sensors or requests from users that need to have some specific information generated from the data collected by the platform. Although all of these requests can be considered valid, there still exists the possibility of a breached device or application call from the user, which could lead to illegal access to private data. The Gatekeeper pattern implies the creation of a support service that acts like a sanitizer, similar to what a firewall does in a classical network topology. It would check incoming requests or data, based on certain specifications and rules implemented by an administrator, verify the authorization of the solicitor and, in case of validity, redirect the request to the appropriate service or discard the request or data, in case of access restriction or unconformity with the packet specifications. In this way, the first layer of control and validation is implemented in the flow of data toward the Cloud platform, thus limiting the attack surface. To ensure that a breach of this module will not affect the background services, it can be decoupled from them and ran in a limited privilege mode, with no direct access to services or storage.

Valet Key

Using a microservice architecture, storage services are commonly stripped of the methods that authenticate or authorize a user or device to access data, these functions being ensured by other services, specialized for these tasks and send just an authorization token. Based on this token, the storage services will allow or deny access to data for the incoming request. Yet, in a Hybrid Cloud profile or in case of failure of some security services, there may exist the need of accessing data without this authorization code. In this case, the Valet key pattern suggests the usage of a key or token, called the Valet key, that the storage service can recognize and accept. Even though the storage service is unaware of the authorization level of the user, the Valet key will provide the necessary information. When generating such a key, it can be configured to include specific information, such as the period of validity, area of storage that can be accessed, types of data that can be accessed or even if permissions are write-only, read-only or read-write. Also, the Valet key can be generated as for a one-operation use, thus it becomes invalid after the user marks as completed an operation that required this key.

6.4.3.2   Data Management Class

Anti-corruption Layer

From a communication perspective, the Gatekeeper design pattern tackles the problem of addressing multiple requests and assessing their validity, but it can block valid requests for not matching the specified rules. This issue can, sometimes, be caused by older devices that do not have the same set of symbols for communicating as newer devices. One possible solution can be the Anti-Corruption Layer pattern that introduces a compatibility module with the intention of isolating older devices from the rest of the IoT system. In this way, requests coming from these subsystems can be translated into requests accepted by every other subsystem, thus by the Gatekeeper component. Even if this pattern can revive the functionalities of older subsystems, it introduces latencies in the entire IoT system, therefore it should be used during a transition period, for integrating a limited number of older devices that will be replaced with newer ones.

Gateway Aggregation

IoT systems are usually characterized by low-power communications, meaning also small amounts of data being transmitted in a single request made by a device. In case of a large amount of data, devices will send two or more data packets toward the Cloud IoT platform and services will be called to parse the incoming data and execute a set of jobs. For security and performance reasons, services will not remain awake and scan every received packet to see if data is intended to them or not, thus a dispatcher will wake them only when needed. If data in a request is marked as continuing by the sender, then the dispatcher could wait for a specific time interval for the remaining sections of data, assemble that data and then call the targeted service. This method is proposed in the Gateway Aggregation pattern, as a means of optimizing the calling intervals of services and also has a security aspect, by minimizing the risk of an attack to be carried by separating malicious instructions in several chunks of data. By placing this module before the Gatekeeper module, a preliminary filtration is ensured and also the performance of the inner components of the Cloud IoT platform is improved.

Queue-Based Load Leveling

Another issue that Cloud IoT platforms need to adapt to is the variability in the number of requests coming in a time interval from both devices and users. This variability can present spikes of traffic that, often, results in congestion when processing those requests. The Queue-based Load Leveling design pattern tries to remedy this situation, by suggesting the implementation of a Message queue. This queue will receive all incoming requests and act as a buffer before messages are then taken over by a service to be processed. The performance of hardware components that the platform is built upon and the way these resources are divided between different services and software modules, remove the limitation of implementing a queue only before a service call. Therefore, other vulnerable components, such as modules responsible for a preliminary parse or filtering of requests, can have a complementary integration of a queue. Also, from a security perspective, the queue transforms a variable flow of requests in a constant one, reducing the risk of a DoS attack on a service.

6.4.3.3   Resource Monitoring Class

Throttling

Cloud systems are basically composed of a multitude of hardware components that provide a considerable quantity of computing power and storage capacity. Even though the performance indicators display seemly sufficient resources for any type of application, these resources are finite and can be easily be consumed if certain criteria are met. To counter such events, Cloud providers have usually implemented many strategies of constraining services in consuming the entire pool of resources. One such strategy is the autoscaling of the resource pool, in which services are defined a certain threshold of resource utilization and, when that threshold is close of being passed, the Cloud system extends the resource pool by allocating new ones from a reserve pool. But, if requests grow quickly, even this strategy can lead to a resource deficit. At this point, the main deficit cannot be considered the hindering of services, but the hindering of their availability, seen as a DoS/DDoS attack from a security perspective. Similar to the auto-scaling strategy, the Throttling design pattern introduces the idea of limiting the resources assigned to services, but when this limit is close of being exceeded, the Cloud system will enter a throttling phase. This phase can be characterized by one of the following options:

- restricting users which called more than a specified number of times the targeted service. This can be done by implementing a monitoring service that notes requests types and their requesters.
- suppressing the execution of lower priority services and reallocate their resources to the service in need. This can be done by implementing a priority scale and a set of rules for assigning them.

Health Endpoint Monitoring

Each IoT system relies on its capacity to safely and uninterruptedly collect information from the underlying myriad of sensors or IoT devices and the Cloud platform. Yet, certain external factors can interrupt these normal flows of information between these two components, either weather elements or attacks executed by malicious users. Given that some connections with the base nodes are established in a power-saving mode, it means that the network components activate only to retrieve commands or data or to send some data or requests. When a configurable time interval expires, there is no possibility of knowing the status of that IoT device between the send or receive cycles. An adaptation of the Health Endpoint Monitoring design pattern can improve the knowledge of IoT device status. Rather than sending requests to Cloud services to verify their integrity and health by the base nodes, as suggested by the design pattern, the IoT nodes can send information packets at regular intervals and not only when sending bulk data. These regular transmissions can improve the security of the IoT systems by having more knowledge of the health of each resource and its availability, so in case of an attack, the perimeter of questionable communications can be rapidly set, so that connections with the remaining devices will not be affected. By adapting the frequency and structure of these information packets, so that they contain minimal data (e.g.,: a simple "OK" message), and the structure of the update packets sent with bulk data (e.g.,: an update packet that contains information related to battery life, processor, and memory usage), the power efficiency can remain constant, even in this new design.

The Health Endpoint Monitoring pattern can be employed in controlling the security of an IoT application that comprises several hardware devices. Taking into consideration the limited hardware capabilities of an IoT device, in terms of secure cryptographic keys storage, an IoT application must implement alternative security solutions in order to make the compromise between security and hardware costs. Given the limited physical security of IoT devices, these modules can be simply relocated and tampered by an attacker. To address this issue, the Health Endpoint Monitoring pattern can be extended to implement a security keep-alive mechanism that ensures the integrity of an IoT system (e.g., devices from a Smart Home). As mentioned in [25], one of the main problems in the IoT device theft is the possibility to impersonate the compromised device. In order to mitigate this attack, a possible extension of the Health Endpoint Monitoring pattern may be used [26]. The security keep-alive mechanism has an initialization phase that consists of a cryptographic key exchange between the newly purchased IoT device and the devices already installed in the system. The key exchange is orchestrated by a user management smartphone application that controls the device enrollment. Thus, the newly purchased device generates an ECC key pair and transmits the public key to the other devices. In turn, each already enrolled device sends its public key to the newly added device. After the key setup is executed, the user management application configures on the newly added device a keep-alive threshold and timeout. Thus, the device will periodically send a challenge packet that has to be signed by at least k-out-of-n peer devices. If this process fails, the device will lock itself and it will not accept commands. The keep-alive packet can be a unicast message sent to each device or it can be a multi-cast message if the transport protocol has this capability (e.g., CoAP). In a multicast scenario, the device will send only one message and will expect at least k responses, this communication solution being more battery friendly. By using the security keep-alive pattern, the IoT system can lock itself in a user-configured timeout period, thus narrowing the attack window for an impersonation attack. The keep-alive extension of the Health Endpoint Monitoring pattern targets autonomous M2M Smart Home applications, where a device or Cloud module can send an actuation command to another device.

Besides these design patterns adapted from the ones proposed by Microsoft for Cloud infrastructures [21, 22], there are still many other designs that can be integrated into the development cycle of a Cloud platform. Some of them are presented in the remaining part of this chapter and will be addressing not only the issues of establishing connections between IoT devices and the Cloud platform and the methods of optimizing data transfers but also some security aspects, such as authentication and data security.

6.4.4 Push Notification Pattern

Cloud-enabled IoT devices are relying on a connection with an external module, like a Cloud application or a third-party management device that is not located in the same LAN. The connection with these Cloud devices must be long-lived, a feature that is against the resource-constrained nature of the IoT devices. A push notification is transmitted using a third-party Cloud service that is not necessarily in the same trust domain as the notification initiator and the targeted IoT device. Considering this reason, the push notification payload needs to be protected by a confidentiality layer or the push notification must bootstrap an end-to-end communication between the targeted IoT device and the notification initiator. The push notification design pattern can also be encountered in the mobile development context as an asynchronous modality of communicating with a smartphone. Thus, the Android SDK and the Brillo SDK (Android Things) are offering APIs for the push notification mechanism. Taking into consideration the smartphone version of the push notification paradigm, this design pattern empowers both the notifications sent to the IoT devices and the notifications sent from the IoT devices to a management system, which can be a smartphone used by the user to manage an IoT network. Because

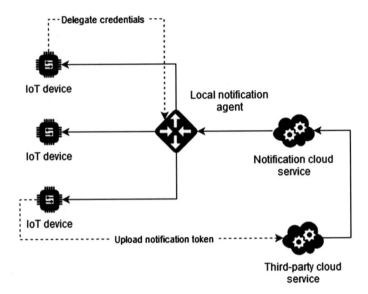

Fig. 6.12 Architecture of the push notification design pattern

the push notification is an asynchronous communication solution, it can be used for real-time synchronization of the IoT devices. The core mechanism of the push notification pattern consists of a software/hardware module maintaining a long-lived connection with a notification Cloud module. The notification Cloud module acts as a message dispatcher for the IoT devices and the notification sender. The local notification module that maintains a connection with the notification Cloud module can be either a process (in the context of an operating system) or a central device (e.g., a central broker or a gateway) that does not have the same resource constraints as the rest of the IoT network. After receiving the notification payload from the Cloud side, the local notification module is in charge of distributing the data into the local network. In Fig. 6.12 is depicted a diagram of the push notification pattern. The push notification pattern consists of the following steps (these generic steps are implemented by the Android system and other similar solutions):

- The Cloud notification system generates a random token and sends it to the notification receiver (IoT device, smartphone).
- The notification receiver uploads the token to the notification sender (e.g., by sending it to a web service).
- The notification sender transmits the token to the notification Cloud, along with a payload, in order to asynchronously notify the IoT device/smartphone.

From the optimization perspective, the local notification module offloads the other modules of maintaining a continuous connection with the notification Cloud. Moreover, this pattern abstracts the push notification logic, allowing the local IoT components or the management smartphone to implement only the payload processing logic. This design pattern is also highly used by the IoT Cloud frameworks (like Kaa IoT) that have an agent on the IoT device communicating with the Cloud platform. Usually, this Cloud agent has data transmission tasks, allowing the IoT device to communicate with both other devices and with management modules, following a virtual network paradigm. A Cloud-enabled IoT device interacts with various modules starting from management devices and ending with data processing web services. Taking into consideration this characteristic, an IoT module must authenticate to a wide range of devices in order to receive asynchronous events. In the IoT context, the push notification pattern can be improved with a security delegation feature, where an IoT device transfers to the local notification module, through a trusted channel, the cryptographic keys/credentials used in the authentication process. Thus, the local notification module can offload the resource-constrained device in executing an authentication process every time an asynchronous event is to be received—the device having the single responsibility of processing the payload. The security delegation task in the push notification context can be implemented only if the local notification module and the third-party Cloud modules (that are communicating with the IoT devices) are sharing a common suite of authentication protocols (e.g., challenge-response). This paradigm can be translated into an IoT operating system context, where the local notification module can be a process communicating with the other processes by using common IPC mechanisms. In this scenario, the client processes can be scheduled only when the local notification process delivers a push notification payload, thus leading to an energy-efficient software structure.

6.4.5 Cloud and Smartphone Management Pattern

A common IoT security scenario is the device enrollment process, which depends on the particular application deployment: industrial, health, Smart City or Smart Home. An IoT device lacks a cryptographic identity and the enrollment process consists of an identity bootstrapping and associating that identity with a resource administrator. Regarding the enrollment process, a widely used security model in the IoT context is the resurrecting duckling, where the IoT device is initially associated with the manufacturer and after that it transitions into the owner trust domain. A widely analyzed IoT enrollment process is the one of Smart Home, where the user purchases the IoT device and has to enroll it into the Smart Home trust domain in a user-friendly manner. The enrollment process allows the IoT module to be associated with an administrator, thus establishing a mutually trusted link that allows the IoT to authenticate the actuation commands received from the management module and the management device to authenticate the IoT sensor acquired data. In the Smart Home context, the smartphone is a widely encountered management device, thus for the Cloud and smartphone management pattern we will focus on an Android-based IoT management application. The core idea of this design pattern is presented in [26]. As software infrastructure, this design pattern uses an IoT Cloud platform that abstracts the hardware architecture: Kaa IoT. The Kaa IoT Cloud platform can generate code in several programming languages, such as C/C++, Java or Objective-C, thus targeting both resource-constrained embedded devices and management devices like Android smartphones. An IoT smartphone management solution consists of two phases:

- The IoT device is enrolled in the Smart Home system, using the HAN connection, this process being user-assisted.
- Each command received by the IoT device is relayed to the user for authorization.

Because, in the Smart Home context, the administrator is a regular user, each security protocol must provide both user-to-device and device-to-service authentication, therefore, in this scenario the device is the Android management module. For the user-to-device and device-to-service authentication, the FIDO (Fast Identity Online) protocol, a passwordless authentication protocol, may be used. FIDO uses asymmetric cryptographic keys and consists of two main phases: the public key and identity (username) registration on the FIDO server side and the actual authentication process following the challenge-response paradigm that uses a signature with the FIDO private key. When the IoT device is installed in the Smart Home, it is imprinted with the manufacturers' cryptographic material: a list of attributes that are used for a CP-ABE-based challenge-response protocol. Thus, after purchasing the device, the user is issued a secret (private) key by the manufacturer using a list of attributes, like the role that can be used when sending commands to the IoT device. In order for the user to begin the IoT device imprinting process, it uses the Android management application that establishes a CoAP connection with the IoT device (CoAP server). The management application initiates a CP-ABE challenge-response protocol: the IoT device generates a nonce, encrypts it with the attributes claimed by the user in the CoAP request and sends it as a CoAP response. The Android application decrypts the nonce with the secret key provided by the manufacturer and makes another CoAP request with the plaintext nonce, to the same CoAP resource. After executing the challenge-response protocol, the IoT device transitions in a state where it accepts other imprinting cryptographic keys, thus being moved into the Smart Home trust domain. For the second cryptographic imprinting stage, the same FIDO protocol may be used, with the IoT device as a FIDO server and the Android device as a FIDO client. Thus, the Android device generates an asymmetric key pair and registers the public key on the IoT device side, as described by the FIDO protocol [27]. As the FIDO specification mentions, the Android managed cryptographic keys are stored on the so-called FIDO authenticator device, which can be implemented by using the Android KeyStore based on TEE (Trusted Execution Environment). For the user to access the FIDO private key, he has to use the Android smartphone fingerprint reader, this process proving the user-to-device authentication. As presented in [26], the challenge-response protocol and the FIDO registration process can be merged into one process, by embedding the challenge-response values as attributes of the FIDO messages. The Smart Home device is connected with the IoT Cloud platform, thus it can receive actuation commands from a Cloud module handling a specific function (e.g., changing the thermostat temperature at a given time, based on the weather prediction). In this scenario, when the IoT device receives the Cloud

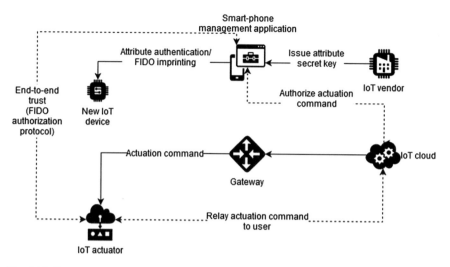

Fig. 6.13 FIDO-based command authorization flow

actuation command, it encodes it in a JSON format and initiates a FIDO authentication session with the Android application, using the IoT Cloud communication platform. The actuation command is transported into the FIDO transaction field and it is displayed to the user by the means of a UI. The user authorizes the command by sending a valid FIDO response, signed with the FIDO private key. The same security mechanism can be applied for the user-initiated commands. By using the FIDO based command authorization method, the IoT device authenticates both the Smart Home owner (by means of the fingerprint mechanism that unlocks the FIDO private key) and Android management device (that sends a valid FIDO response, thus proving the private key ownership). Fig. 6.13 is presented the FIDO-based command authorization flow. This design pattern protects the Smart Home entities (IoT and management devices) against a Cloud platform attack, that may send malicious actuation commands. A common Smart Home communication method consists of a broadcast message sent from the Cloud module to the Smart Home devices. By applying this design pattern, the initial broadcast message is followed by unicast messages that consist of the FIDO authentication steps, finished with the actuation command execution.

6.4.6   Cloud-Assisted Network Access Pattern

Smart Home is one of the most encountered IoT applications, where a user manages several IoT devices communicating with a third-party Cloud. The Cloud module permits the communication between the IoT devices and the management module, which can be a smartphone or a web application. The Smart Home context is a

dynamic environment, where transient devices can establish ad-hoc connections with
the IoT modules located in the Smart Home. In this scenario, the user must authorize
a temporary connection, where the transient IoT device can use the Smart Home
network services or the Cloud application services. The Cloud-assisted network
access pattern allows the user to asynchronously authorize a third-party device, using
a smartphone as a management device. This design pattern allows a user to authorize the transient device even if the management device (smartphone) is located in
another network (communicates with the HAN via an Internet connection). In this
scenario, the transient IoT device can be authenticated by using a Layer 2 protocol
like 802.1X (EAP) or it can be authenticated at the application level. In a classical
Wi-Fi enterprise network, an 802.1X architecture uses a RADIUS server for handling
the actual authentication process, the gateway (NAS) being a module that relays the
packets between the supplicant and the RADIUS. In contrast, a HAN cannot apply
strict security policies and device credentials/attributes registration on the RADIUS
server side, the Smart Home applications having an important functional requirement: user-experience. Taking the aforementioned requirement into consideration,
this design pattern allows the user to actively participate in the transient IoT device
authentication process, thus achieving a user-friendly solution providing the desired
security level. The Cloud-assisted network access pattern adds another component
to an existing 802.1X authentication infrastructure that may exist in a HAN based
on off-the-shelf networking hardware. To implement this design pattern, a custom
backend RADIUS module can be used: open-source RADIUS servers, like FreeRA-
DIUS, can have multiple backend modules that can be used as credentials repository (e.g., a MySQL/MariaDB database) or can implement a custom authentication
protocol. The FreeRADIUS solution also has a REST backend module (rlm_rest),
that can transmit to a configured web server the credentials (username/password)
delivered by the IoT device via EAP. After the web server receives these credentials, it relays them to the user management device (smartphone) by using the push
notification design pattern. When the users' smartphone receives the notification,
the management application displays a UI interface that presents to the user, the IoT
device credentials and the user can allow access for a certain amount of time, by
also using UI elements. After receiving the user response, the webserver caches the
authentication status for a certain amount of time (configured by the user) and serves
this authentication status to the FreeRADIUS server without having to contact the
user. In this scenario, a trust relationship must be established between the user and
the webserver and every time the user authorizes an IoT device, he must authenticate to the smartphone and the smartphone must authenticate to the web server, thus
providing both user-to-device and device-to-service authentication. For the user-
to-device and device-to-service authentication, the FIDO protocol may be used in
a manner similar to the one described in the Cloud and smartphone management
design pattern. The Cloud assisted network access design pattern is presented in
Fig. 6.14. By implementing this design pattern, the user can easily control the IoT
device enrollment in the Smart Home context, using a security solution that satisfies
the user-experience requirement. As an alternative, the IoT modules not supporting
the 802.1X protocol, can be authenticated using the MAC authentication bypass

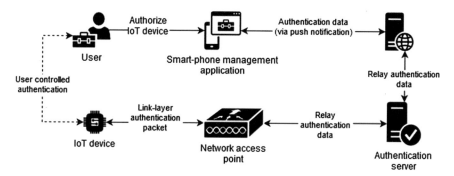

Fig. 6.14 Architecture of the cloud-assisted network access design pattern

(MAB) mechanism, which notifies users, via a push notification, that a new MAC address tries to access the HAN. The MAB solution can be considered a security compromise, due to its vulnerability to MAC address spoofing, while also allowing legacy devices to use the HAN services.

6.5   Discussion and Conclusion

The IoT security landscape is becoming more and more diverse, with the resource-constrained embedded IoT devices creating new attack surfaces. The core element of the IoT architectures is represented by the IoT network, viewed as a closely bounded collection of components that can be either an IoT endpoint network or an IoT Cloud platform. When architecting the IoT network application, multiple aspects have to be taken into consideration, starting with addressing security issues and ending with satisfying the functional and performance requirements. Even though the IoT network applications can be complex and in a wide range of flavors, by targeting multiple application domains, from the security perspective there a series of recurrent issues that appear in each IoT network/Cloud application. Our work encompasses a series of architectural security design patterns that can help in accelerating the design and the development of an IoT solution. The presented design patterns address security issues regarding IoT lightweight communication protocols and issues regarding the integration of IoT devices with Cloud platforms and smartphone devices. Taking into consideration the complex structure of an IoT application, it is a common practice to combine several design patterns in order to obtain the desired structure and results. From the security perspective, using a design pattern brings several advantages, such as: reducing the time/resources allocated for security development and leaving more time for developing the application business.

   The presented design patterns are focused on software exclusive solutions that solve a particular security problem. This characteristic allows the use of design patterns on low-end IoT devices, without having to make an assumption regarding the hardware capabilities, like the existence of TPM (Trusted Platform Module) to store and execute the cryptographic operations. Regarding the software platform required to implement the presented design patterns, we would like to stress the necessity of using an operating system for the majority of them. Even though IoT are resource-constrained devices, it is a common practice to run a lightweight operating system or an RTOS, such as embedded Linux (Yocto), Zephyr or VxWorks. From the cost perspective, the edge of an IoT network is populated with low-end heterogeneous devices (difference in hardware capabilities and different vendors), thus a hardware-agnostic security design pattern gives the solution architects flexibility in terms of choosing the hardware.

A particular IoT implementation characteristic is the deployment context. For instance, an M2M scenario is more flexible in terms of the chosen security solution, being able to implement complex authentication/authorization algorithms, limited only by the communication protocol and by the device capabilities. IoT deployment contexts that involve the user interaction must take into consideration the user-experience and must implement simple, controllable, and transparent security solutions. The presented design patterns are taking into consideration the two aforementioned characteristics and stress the importance of using a user-controllable IoT management device, like a smartphone.

The need of integrating software components design patterns has not been more visible than in these past few years, given the increasing number of attacks that exploited vulnerabilities not of various concepts or standards, but their software implementation. Among the most known of these vulnerabilities, we can mention Heartbleed (exploit for the SSL protocol), Shellshock (exploit for the Unix Bash shell), the mix of exploits for the TLS protocol (POODLE, BREACH, HEIST), ROCA (exploit in the implementation of RSA in Infineon chips), massive cyberattacks, such as the DYN cyber attack (series of DDoS attacks that targeted the systems of Dyn, a DNS provider), or classic software implementation exploits, such as SQL Injection or Buffer Overflow. These are only a part of the cases in which unverified methods of implementation can raise issues, even in a seemly perfect system. Also, these attacks are an example of what design patterns try to overcome, by providing software implementation methodologies based on previous experience and multiple tests. Many of the previously mentioned exploits have been fixed by replacing faulty software modules with new ones that have been properly tested and developed using proven implementation designs. Thus, design patterns can sometimes make the difference between a secure or insecure environment, a fast or slow communication link or data processing module. Moreover, in a resource-constrained environment as IoT systems, these different settings can have a big impact on the overall performance of the system, therefore design patterns can provide the means of adapting to various scenarios.

An important future direction is completing the presented architectural design patterns with a development framework that implements these security solutions as software libraries. The development frameworks are highly popular, starting with well-known web frameworks like Spring and ending with IoT frameworks, like

Eclipse Kura. A security design pattern framework that addresses several communication protocol issues and targets both IoT and management devices would complete the already existing functional development frameworks.

References

1. Willemsen B, Mahdi D (2018) Preserve privacy when initiating your IoT strategy. Gartner Report, ID: G00317400
2. Bandyopadhyay D, Sen J (2011) Internet of things: applications and challenges in technology and standardization. J Wirel Pers Commun (Springer) 58(1):49–69. https://doi.org/10.1007/s11277-011-0288-5
3. Miorandi D, Sicari S, Pellegrini F, Chlamtac I (2012) Internet of things: vision, applications and research challenges. J Ad Hoc Netw (Elsevier) 10(7):1497–1516. https://doi.org/10.1016/j.adhoc.2012.02.016
4. Zhang Z et al (2014) IoT security: ongoing challenges and research opportunities. IEEE 7th international conference on service-oriented computing and applications (SOCA), pp 230–234. https://doi.org/10.1109/SOCA.2014.58
5. Microsoft (2016) Securing your internet of things from the ground up. Whitepaper
6. Shafi Q (2012) Cyber physical systems security: a brief survey. In: 12th international conference on computational science and its applications, pp 146–150. https://doi.org/10.1109/ICCSA.2012.36
7. Kandukuri BR, Ramakrishna Paturi V, Rakshit A (2009) Cloud security issues. In: IEEE international conference on services computing, Bangalore, pp 517–520. https://doi.org/10.1109/SCC.2009.84
8. Aljawarneh SA, Alawneh A, Jaradat R (2017) Cloud security engineering: early stages of SDLC. J Future Gener Comput Syst (Elsevier) 74:385–392. https://doi.org/10.1016/j.future.2016.10.005
9. Vogel B, Gkouskos D (2017) An open architecture approach: towards common design principles for an IoT architecture. In: Proceedings of the 11th European conference on software architecture: companion proceedings. ACM, pp 85–88. https://doi.org/10.1145/3129790.3129793
10. Bonomi F, Milito R, Zhu J, Addepalli S (2012) Fog computing and its role in the internet of things. In: Proceedings of the first edition of the MCC workshop on mobile cloud computing. ACM, pp 13–16
11. Anitta V, Fincy F, Ayyappadas PS (2015) Security aspects in 6 low-pan networks: a study. IOSR J Electron Commun Eng (IOSR-JECE) 8–12
12. Jones M, Bradley J, Sakimura N (2015) JSON web token (JWT). RFC 7519. https://doi.org/10.17487/RFC7519
13. International Organization for Standardization (2016) Information technology—message queuing telemetry transport (MQTT) v3.1.1. International Standard ISO/IEC 20922:2016
14. Shelby Z, Hartke K, Bormann C (2014) The constrained application protocol (CoAP). RFC 7252 https://doi.org/10.17487/RFC7252
15. Lucas R (2017) DTLS multicast. Internet Draft. https://tools.ietf.org/html/draft-lucas-dtls-multicast-00
16. Hardjono T, Maler E, Machulak M, Catalano D (2015) User-managed access (UMA) profile of OAuth 2.0. Specification created by the User-Managed Access Group of the Kantara Initiative
17. Tschofenig H, Maler E, Wahlstroem E, Erdtman S (2015) Authentication and authorization for constrained environments using OAuth and UMA. Internet Draft. https://tools.ietf.org/html/draft-maler-ace-oauth-uma-00

18. Chifor B-C, Bica I, Patriciu V-V (2017) Mitigating DoS attacks in publish-subscribe IoT networks. In: 9th international conference on electronics, computers and artificial intelligence (ECAI), pp 1–6. https://doi.org/10.1109/ECAI.2017.8166463
19. Cozzolino V et al (2017) Enabling fine-grained edge offloading for IoT. In: Proceedings of the SIGCOMM posters and demos. ACM, pp 124–126. https://doi.org/10.1145/3123878.3132009
20. Grieves M (2014) Digital twin: manufacturing excellence through virtual factory replication. Whitepaper
21. Homer A et al (2014) Cloud design patterns: prescriptive architecture guidance for cloud applications (Microsoft patterns & practices). Microsoft, ISBN: 978-1-62114-036-8
22. Wilder B (2012) Cloud architecture patterns: using Microsoft azure. O'Reilly, 1st ed. ISBN: 978-1449319779
23. Ghazizadeh E, Zamani M, Manan J-A, Pashang A (2013) A survey on security issues of federated identity in the cloud computing. In: Proceedings of the 4th IEEE international conference on cloud computing technology and science. https://doi.org/10.1109/CloudCom.2012.6427513
24. Chadwick DW et al (2014) Adding federated identity management to OpenStack. J Grid Comput (Springer) 12(1): 3–27, Online ISSN: 1572-9184
25. Stajano F (2011) Pico: no more passwords! In: Security protocols 2011: security protocols XIX. Lecture notes in computer science, vol 7114. Springer, pp 49–81
26. Chifor B-C, Bica I, Patriciu V-V, Pop F (2018) A security authorization scheme for smart home internet of things devices. J Futur Gener Comput Syst (Elsevier) 86:740–749. https://doi.org/10.1016/j.future.2017.05.048
27. Srinivas S, Kemp J, FIDO Alliance (2017) FIDO UAF architectural overview. FIDO Alliance Proposed Standard

Chapter 7
Cloud-Based mHealth Streaming IoT Processing

Marjan Gusev

Abstract Advances in IoT technology have resulted in extensive use in mHealth applications. Especially, the use of wearable devices and smart home solutions has boosted the development of mHealth applications, even to domains where no one has dreamed for. From the other side, cloud-based computing is the only alternative to enable real-time remote monitoring, information exchange with the outer world, and access to enormous Big data applications. This is a real motivation for customers, since they crave a certain level of security that someone is monitoring their health with the latest technology and at the same time, it gives them a chance to prevent serious health damages, without lacking the freedom to operate in their daily activities and home environment. A lot of challenges are met to realize such a system, and this paper presents an overview of architectural approaches and organizational methods to realize a cloud-based mHealth IoT application that will cope with the Big data concept of incoming data streams with high velocity and volume.

Keywords mHealth · IoT · Cloud computing · Edge computing · Dew computing · Implantable mhealth devices · Wearable mhealth devices · Smart pocket devices · Monitoring mHealth center

7.1 Introduction

The mobile-devices assisted health care and medical applications are expected to create the next breakthrough in demand for smartphones and mobile services [1, 3]. It improves the patient's lives, especially in the elderly, disabled, and chronically ill. The available technology, and the need to provide direct access to health services, regardless of time and place. Heterogeneous user groups are attracted by mHealth solutions, unifying altogether doctors, nurses, patients, or even healthy people [4].

M. Gusev (✉)
Faculty of Computer Science and Engineering, Ss. Cyril and Methodius University, Skopje, Macedonia
e-mail: marjan.gushev@finki.ukim.mk

© Springer Nature Switzerland AG 2021
F. Pop and G. Neagu (eds.), *Big Data Platforms and Applications*,
Computer Communications and Networks,
https://doi.org/10.1007/978-3-030-38836-2_7

Mobile phones and/or smartphones target a wide variety of purposes [10], including personalized assistants for losing weight loss, practice a particular diet and physical activity, treatment, and disease management. In addition, the key factor that contributes to the development of mHealth can be remote monitoring [14].

Latest market research studies [16] identify key trends to support this claim, setting a focus on smartphone penetration as the main development driver, and the ability to customize and personalize dedicated applications, that will be always present with people and provide intelligent messages to support a healthier lifestyle [19]. The estimated mHealth market for 2019 [16] reaches 37bn USD.

Although there is a huge variety of mHealth applications targeting prevention and healthy lifestyle, obtaining information on a specific healthcare service or facility, remote monitoring, diagnosis, personalized assistance in disease treatment and management, we will target only those mHealth applications that use the Internet of Things (IoT) and stream data to the smartphone.

IoT devices provide a huge variety of wearables and smart home solutions that can be efficiently used in the mHealth, such as wearable sensing devices used for monitoring a certain health-related parameter, or smart home devices that support a better visualization, communication, and information exchange with the outer world. However, not all of these solutions stream data and belong to the world of Big data applications.

There are a lot of issues in the realization of streaming IoT solutions that need Big data processing applications, especially in the mHealth area, starting from how to reduce the throughput demands, how to organize storage and processing of data input with high volumes and velocity. This paper gives an overview of architectural and organizational approaches to cope with these demands.

The remainder of the paper follows the next structure. Section 7.2 defines mHealth applications and categorizes them according to several criteria. The application of IoT devices in mHealth solutions is analyzed in Sect. 7.3. Section 7.4 analyzes the cloud-based solutions for these applications and Sect. 7.5 elaborates the issues in streaming IoT applications to realize Big data applications. The architectural principles and organizational approaches for analyzed applications are presented in Sect. 7.6. Comparison of these approaches and benefits of applying these architectural and organizational approaches are discussed in Sect. 7.7. Finally, conclusions and future directions are given in Sect. 7.8.

7.2 Overview of Underlying Technology for mHealth Solutions

An area based on the provision of medicine and health care by mobile devices is referred as mHealth. In this section, we analyze the underlying technology and role of mobile devices in delivery of mHealth solutions.

Although a mobile device is usually associated with a device used for communication purposes with wireless radio technology and is mobile in respect to the physical environment, the mHealth refers to a wider range of computing and communication devices, including personal digital assistants, tablets, laptops, smartphones, smartwatches, wearable devices, and even other forms of personal or home computing devices using electronic technologies with an Internet access.

The underlying information and communication technology hardware that provide infrastructure for mHealth applications include computer networking, mobile operators networks, servers, and other computers, communications satellite, etc. The enabling technologies that resulted with development advances of mHealth are summarized in Table 7.1.

For users, mHealth mainly means delivery of health services and information, and collection of health-related information from various sources including nearby sensors or various shared Internet resources. Users are not limited to patients, but actively involve medical staff, practitioners, and researchers.

The mHealth applications can offer the users real-time monitoring of health-related vital signs, and direct or indirect healthcare provision, as specified in Table 7.2.

The identified functions to be realized by the software solutions include data collection, transport, storage, processing, and result retrieval. These are summarized in Table 7.3.

Data collected by various sources is usually raw data that needs to be processed to deliver meaningful results. The main question a designer of mHealth solution is concerned is where the processing should be realized, should it be on the place where the user is located, or can it be realized at a remote server with higher computing capabilities and where one can gather various information from different sources in

Table 7.1 Technologies that enabled advanced development of mHealth

Technology	Description
Medical devices	A medical device is hardware and/or software used to diagnose, prevent, or treat a health condition or disease without biological or chemical action on any part of the body. Examples include supporting devices, such as wheelchairs, surgical lasers, orthopedic pins; sensing devices, such as heart rate sensors, body temperature or blood pressure measurements, or controlling devices, such as pacemakers, insulin pumps, etc.
Communications	Technologies that support sending short textual messages (SMS), talking to a medical expert or caregiver by conventional mobile operator's networks, exchanging information on social networks
Networking	Data and computer networking infrastructure and Internet are essential for the provision of mHealth, organization of cloud servers and delivery of applications and services, browsing information on Internet about the treatment of a medical condition, or any related health information
Software	Applications that enable processing of raw data, detection of health issues, diagnosis of medical status, intelligent messaging, reminders, and treatment control

Table 7.2 Categories of mHealth applications categorized by nature of their use

Category	Description
Wellness	Applications that provide practical information to measure physical activity, strengthen the immune system, protect from seasonal diseases, manage disease treatment, use dietary supplements, and healthy eating habits
Prevention	Personalized application customized to help users control their health status with information about allergy triggers, including pollen count, pollution, and weather; to access web-based patient information, to control the intake of medicaments, based on prescribed therapy and measured indicators or track health-related parameters to be shared with medical experts and care providers
Monitoring	Applications that provide a remote monitoring solution to users, care providers, and doctors; technology-based diagnosis support and remote patient surveillance, intelligent autodetect features that alert on critical health status, and advanced point-of-care diagnostic capabilities for early detection of disease progression and enhanced remote management of treatment or medical condition
Awareness	Apps that educate users, raise awareness, or provide helpline with medical expertise and explanations

Table 7.3 Functions realized in mHealth applications

Category	Description
Data collection	Data is collected by nearby sensing devices, internal sensors in the device, or information collected by other remote sources on local area network or Internet
Data processing	Collected raw data is processed by sophisticated algorithms in the device matching the energy availability constraints and processing demands
Data presentation	Available display of meaningful results is essential in mHealth applications to communicate with the user, by a signaling LED or small screen with textual or visual presentation
Information exchange	Data processing usually needs exchange of information with other systems on Internet, in order to get information about the surrounding environment or to offload data and computation to a remote server

realization of the medical diagnosis and identification of health problem initiators and healthcare treatment.

This analysis tackles the architectural concept of building the mHealth application. Table 7.4 gives an overview of various approaches a designer of mHealth solution is implementing.

Table 7.4 Architectural concepts to build of mHealth applications

Category	Description
Communication only	Applications that include the exchange of text messages (SMS), direct phone conversations, or social network interaction
Standalone apps	Smartphone, smartwatch or smart home device applications that are used for indication of health-related measurements obtained by sensors and processed internally in the device, usually personalized and customized to the user
Cloud apps	Services and applications hosted on cloud servers and exchange information with the outer world to make a better and intelligent solution besides the internally collected information
Dew apps	Applications on smartphones, smartwatches or other smart home devices used as intermediate processing units that can process internally collected data and work without or with and the Internet to access and exchange information from the outer world

7.3   Overview of IoT mHealth Solutions

A network of various devices, including sensors and actuators that interact seamlessly with the environment around us [2], which communicate via the Internet with each other is called IoT.

An IoT sensor converts the sensed environmental signals to data and provides them to other devices. The other devices process them to obtain meaningful results about a specific physical, chemical, or biological measure. When applied to mHealth, these sensors usually process health-related indicators. The basic functions such an IoT sensor provides are sensing, transforming the signal into digital data, and transmitting digital data to a nearby device [6].

An IoT actuator is a device that accepts information that will trigger an action on the device. Examples In mHealth applications include a pacemaker that can trigger a heartbeat whenever it detects a longer pause, or an insulin pump that can insert a specific dose of insulin according to the measured glucose level.

An IoT controller contains both the IoT sensor and actuator in one device. However, IoT sensors and actuators may be realized as autonomous separate devices, or an mHealth application may communicate to several IoT sensors and actuators. Generally analyzing the architectural building concepts, the communication between the IoT sensor and controller devices can be realized by different communication technologies, including a direct cable connection, Bluetooth, or another personal area or LAN connection.

IoT devices can appear in different forms and a simple categorization is presented in Table 7.5.

Table 7.5 IoT mHealth device categorization

Category	Description
Implantable devices	Devices that are implanted in the human's body, such as a pacemaker, insulin pump, or similar, that do not produce any biological and chemical reaction directly, but trigger an electrical impulse or insert a certain quantity of biomaterial
Wearable devices	Devices worn on the human body, such as patched sensors to sense some health-related indicators
Smart pocket devices	Smartwatches, smartphones or any other similar device that is not patched to the human body but is usually held in the nearby pocket close to the human
Smart home devices	Communication boxes, gateways, interactive TVs, or similar devices that can establish communication with the user and other sources on the Internet
Tablets, laptops	Computing devices that are mobile and wirelessly connected and can be used to establish communication with Internet and other users

7.4 Cloud-Based Architectures

Cloud servers are usually used to offload data and computations by the IoT devices, or to share information with the outside world.

A typical cloud-based mHealth architecture uses the Internet and Web services, where doctors and patients can interact, not only just to access the same medical record anytime and anywhere by any computing device, but also to enable real-time monitoring. Figure 7.1 presents a typical cloud-based mHealth solution. The analyzed mHealth cloud service interacts with other services, to exchange relevant health-related information and update the internal health record. It interacts with the user via a mobile application and with the doctor and/or caregiver via a web application.

The cloud mHealth service interacts with the mHealth application on the smartphone to enable two functional modes, one with the existence of the Internet, where all data is exchanged with the cloud-based service, and the other as a standalone application, where the smartphone functions independently.

The first mode means realization of the *edge computing* architectural solution [6, 11–13], where an intermediate device (smartphone) is located between the user and cloud server. This device is on the edge of the Internet network and enables the proper functioning of the mHealth solution.

The existence of the second mode in addition to the first mode means that it is the realization of the *dew computing* solution [9, 15, 17, 18], where the device can be on the edge of the Internet network, but it can also operate independently as a standalone application.

Analyzing these architectural concepts, this cloud-based solution does not involve any IoT device. Figure 7.2 presents such an architecture, where several IoT devices

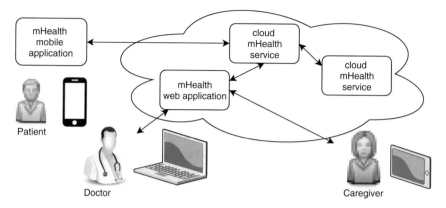

Fig. 7.1 A typical cloud-based mHealth application consists of a mobile application, accessible by the patient; and a web application accessible by a computing device that belongs to the doctor or caregiver

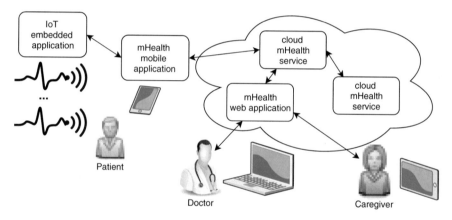

Fig. 7.2 A cloud-based mHealth IoT solution integrates several IoT devices via a mobile application. Each IoT device runs embedded IoT software to sense data and transmit to the mHealth mobile application running on a smartphone. The cloud contains several cloud mHealth services and a web application to interact with the doctor's or caregiver's computing device

are added to the previous architectural design to sense health-related indicators and transmit data to the mHealth application on the smartphone.

The presented cloud-based mHealth IoT solution can be realized by the previously described architectural concepts: the *edge computing concept*, where the smartphone needs the Internet to work, or by the *dew computing* architectural concept, where the smartphone functions as a standalone application, that can exchange data with cloud-based services if there is Internet to synchronize and collaborate with other services.

7.5 Issues for Streaming mHealth IoT Solutions

The presented cloud-based mHealth IoT solution can not be efficient in all working examples. The problems start to arise when the incoming data starts to increase volume and velocity. IoT-based mHealth applications can be classified according to the volume and velocity they arrive from the IoT device in the smartphone or cloud server.

Table 7.6 presents a simple classification base on streaming IoT capabilities. Values defined as a low, medium, or high volume and velocity may differ through time and depend on the capabilities of the associated technology. In the context of the current available technology, communication capacity is measured by velocity of incoming data and can be expressed as KB/s or MB/s.

For example, several samples per second are associated to be less than 0.1 KB/s; hundreds of samples per sec is less than 5 KB/s; and thousands of samples per sec means more than 100 KB/s. Similar to this explanation, volumes can be measured on a monthly basis, or per year. Following the amount of data velocity, the generated volume is low if it is less than 260 MB per month, medium if it is less than 13 GB per month, and high if it is more than 250 GB per month.

According to the Big data definition using 5 Vs (volume, velocity, value, variety, and veracity), in Table 7.6 we have only analyzed the volume and velocity to define these IoT devices as devices that stream data to the smartphone or cloud server. So, the last two categories associated as streaming and intensive can be classified to be processed by *Big data* applications.

Let's analyze the sources where the problems come from. IoT devices need to be small in order to be integrated into implantable, wearable, or pocket type IoT devices. Their size enforces the designer of such an IoT device to build a very small battery with limited capacity.

Issues related to IoT devices are mainly addressing the energy and size that indirectly dictate processing, storage, and communication capabilities. However, user requirements are opposed to the technical capabilities. Although one likes an implantable or wearable device to be a device that realizes complex operations and

Table 7.6 Types of mHealth solutions categorized by streaming IoT capabilities

Classes	Velocity	Volume	Examples
Stationary	Several samples per day	Low	Body temp, blood pressure, glucose, etc.
Frequent	Several samples per sec	Low	Beat rate, physical motion, etc.
Streaming	Hundreds samples per sec	Medium	ECG, EEG, etc.
Intensive	Thousands samples per sec	High	Continuous video

stores a lot of data, it is usually accompanied by a very small battery that demands integration of small processing and communication capabilities. In addition, the users would like that these implantable or wearable devices to last as much as possible, which is also contradictory to the size constraints.

Increased energy availability is found in smartwatches and smartphones that can allow more processing, storage, and communication. Smart home and personal computing devices can use uninterrupted power supply and in this case can have greater processing, storage, and communication capabilities.

A designer of an IoT mHealth solution must precisely analyze energy consumption needs and make a comprise with processing, storage, and communication demands. Especially, this is an important requirement in mHealth applications with intensive streaming data.

The compromise between demands to last longer time and realize complex operations, form one side, and limited energy supply, from another side are explained by specialized architectural designs and organizational approaches.

7.6 Architectures for Streaming mHealth IoT Solutions

The technical problem in the realization of a cloud-based mHealth solution based on streaming IoT data is how to organize and process data coming with high volume and velocity. Currently, a lot of projects and research is conducted to establish an efficient solution.

In order to save energy consumption, the small-sized IoT device needs to perform as few operations as possible, and store no data. This means that it should offload all data. The question to choose where to offload is once again constrained by the small battery and the only alternative is to use personal area networking protocols and offload data to a nearby device. In case a wired connection is not possible, the only alternative is the use of Bluetooth, or similar radio transmission technology, or even ultrasound or other relevant technology that can be used on the nearby smartphone.

The smartphone will take over all relevant computations and takes the role to receive data, temporarily store, process them as much as possible, and finally transmit information to the cloud server. In addition, the smartphone is used to interact with the user and control the execution of the embedded software on the IoT device, mobile mHealth application hosted on the smartphone, and communicate to the cloud service to exchange information. The cloud server will receive all data, realize comprehensive computations, and exchange meaningful results and information with other services.

We present an architecture of a solution for streaming IoT and cloud-based mHealth application in Fig. 7.3.

This mHealth IoT design assumes that the mHealth cloud service will require several servers, including a file server to collect receiving files and data by the patient's smartphone, then main server to act as a web application that interacts with external

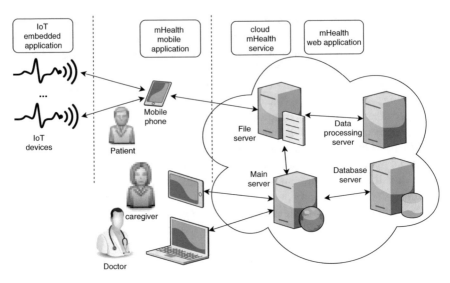

Fig. 7.3 The architecture of a streaming IoT mHealth application integrates an IoT level with embedded applications in wearable or other IoT sensors; an mHealth user interface level with a mobile application on the smartphone and a web application on mobile computing device; and cloud service level with mHealth services and web application hosted on separate servers, including, at least the file, data processing, main and database servers

users, data processing server that processes incoming data to find a meaningful result, and database server that stores relevant information to be used by the main server.

The high-level concept behind a streaming IoT mHealth solution is based on a light wearable sensor that can be patched easily on a human's body or implantable device, or any other nearby small sized IoT device. It connects wirelessly to a mobile device application (smartphone, smartwatch, tablet, or other similar device), which displays the monitored data in a user-friendly manner. The sensor can be worn with no restrictions on physical movement for however long a patient wishes and even in the unlikeliest of places, like the shower.

The sensor plays the role of amplifying the sensed signal and filter the surrounding noise prior to making the conversion to raw data samples. The samples are buffered and organized in data chunks to be transferred by a personal area network, such as Bluetooth Low Energy. The nearby smartphone, tablet, or other mobile device plays a role of an intermediate device, that is a dew server. A dew server receives the signal, and then it continues to process, store, visualize, and transmit data to the cloud server, or works independently as a standalone system.

A user interface enables the patient to control the sensor (pair, connect, disconnect), analyze the collected data chunks, form continuous data streams as separate files to transmit to the server via WiFi, 3G/4G or other radio connection. The user can monitor locally and record an event if they feel a heart-related problem, such as dizziness or another health issues.

The cloud server receives these files and stores them with a corresponding file manager service. The data processor performs feature extraction, analysis, and diagnosis to create advanced alerts and update the database for dashboard information, which is presented to the doctor along with extended visualization features in order to make a proper decision for diagnosis and intervention.

The technological idea for the service is based on an automated monitoring software agent, which alerts in the case of detected health disorders. Currently, diagnosis solutions can be found at hospitals but are not affordable to everyone. The cloud center can use sophisticated algorithms, artificial intelligence, and parallel processing methods to enable simultaneous processing of multiple data streams with high volumes and speed. This technology consists of modern trends for organizing a cloud computing mHealth solution and mobile application.

7.7 Discusion

In this section, we compare the presented architectural concepts and organizational approaches with other existing approaches. This comparison is followed by an analysis of the benefits of such a solution, and a use case of a future monitoring that will be based on mHealth solution and wearable streaming IoT devices.

7.7.1 Comparison of Architectural Concepts

Producers of modern medical devices mainly market their products as high technology solutions available for hospitals and caregiving centers. However, there are a very small number of devices that offer solutions for patients as end-users. mHealth is the real platform for this type of solutions. Currently, the only offer of mHealth solutions are addressing access to shared health records on the cloud, or personal assistants that give advice on prescribed medical therapy, or how to enjoy a healthier lifestyle by analyzing the physical activities and food intake.

Direct implementation of a cloud-based solution is not possible, due to several constraints. A smartphone does not have the possibility to integrate all sensors that sense health-related parameters and need nearby IoT devices. So the complete architectural design needs at least three layers: IoT device software, smartphone mobile application, and cloud services. As we have discussed streaming data needs changes to the design.

In this paper, we analyze mHeath solutions that are used for real-time health monitoring, as they represent a real technology breakthrough and different philosophies used by main producers of medical devices. The analyzed issues are mainly targeting streaming data coming from accompanied IoT devices. In addition, the mHealth mobile application is linked to a cloud service in order to offload data and information for further processing and storage.

Our design uses a sophisticated edge computing concept with one or more intermediate edge devices and servers [6]. The idea is to bring the computing closer to the user and eliminate the long delays introduced by wide area networks. It functions as a kind of locating a cache between the processor and memory. The smaller the cache memory is, the faster it loads and stores data. In our case, the smaller the computing device it consumes less energy to receive data. As we need more processing and storage, we need to offload data to a more powerful edge device or server. According to this philosophy, a smartphone is close enough to the IoT device, and it is an edge device that can sufficiently process some operations, temporarily store data, and communicate with the cloud server.

This solution, also, follows the *dew computing* concept [5], since the edge device (smartphone) can also function independently as a standalone application, and the mHealth solution will not depend on Internet availability. The device can synchronize data and collaborate by exchanging information with nearby devices.

The mHeath solution using streaming IoT data can also be organized using the serverless real-time data analytics platform [8]. These solutions do not use servers for data analytics processing, but schedule processing and storage requirements among available devices.

7.7.2 Benefits

These mHealth solutions address various issues appearing in health services. The reach of mHealth solutions includes those with chronic diseases related to their lifestyle, and decreases the high costs of existing health services. It also enables patients and families to practice a better healthier lifestyle, self-care and handle their own healthcare treatments using the benefits of the new technology to have an access to healthcare services anytime and anywhere.

mHealth solutions can help healthcare providers deliver better and more efficient healthcare regardless of time and location. When wearable devices are used, then mHealth benefits even more, since it allows the users to freely move and perform daily activities without any obstacles. Their health will be monitored and alerts triggered in the case of a health threatening situation.

In addition, mHealth gives the patients another very important feature, they do not need to schedule and queue for doctor's appointments in order to have regular check-ups unless the system alerts them of abnormality. This will make people feel more secure and, at the same time give them more time to spend on things they love.

Ageing can expose discrimination knowing that attention to it is vulnerable for the elderly. This kind of mHealth solutions will give the elderly a better opportunity to live independently and feel secure, independent of any sex or gender divergence. It can be used by people of any social or political factor and biological characteristics.

The software solution used for the available devices is mainly based on capturing and collecting monitored data, storing, and transmitting at a later stage to the medical expert. Most of the existing software products are based on postponed analysis, for example, after a 24 h measurement is realized, and do not address real-time monitoring.

Such real-time monitoring which streams data from an IoT mHealth solution can implement triggering an alert when an abnormal health condition is detected for non-hospitalized users.

7.7.3 Use Case: A Monitoring Center Based on Streaming IoT mHealth Solutions

The final goal of providing streaming mHealth IoT solutions will be to establish a monitoring center based on online remote monitoring and predictive intelligence for the early detection of any health disorder. The monitoring center will be based on a fully operational and scalable system to offer such a service to a wider population.

The service will be based on an automated detection of health disorders, and once the alert is triggered, it is sent to the doctor to confirm the health status. An example of such a system is used for real-time ECG monitoring and alerting of detected arrhythmia [7]. The system will send out an emotionally intelligent message to the user so as not to generate stress, but to instead give gentle advice and tips for relaxation on how to treat the detected health disorder. For example, in an extreme case, a doctor in the monitoring center can contact the patient to make a better decision or may call an emergency ambulance to prevent serious health damages.

A very important advantage of such a system is to offer health monitoring as a service affordable to the general public, not just to hospitalized patients. The remote monitoring needs to be accessible both to the patient and caregiver, or their family doctor. The monitored data is sent to the doctor for reference and personalized medical advice is available at premium service.

Automatic recognition of abnormality and real-time alerting is of great importance and gives a great advantage to existing monitors, especially in cases when the user does not directly feel pain and triggers an event. The presented mHealth solution for remote health care without hospitalization addresses the following user needs: quick detection, quick intervention, and feel secure knowing that someone takes care of the heart condition.

7.8 Conclusion

The architecture of an mHealth solution using streaming IoT devices was presented along with organizational approaches. The presented Big data solution copes with extensive data coming with high velocity and volume. A use case was presented of a cloud-based monitoring center that can accept, process, and respond in real time to the demands of real-time monitoring and alerting in case a dangerous or life-threatening health situation is detected.

The architecture is based on a wearable, implantable or small-sized IoT device, that senses a certain signal and transfers data samples to the nearby smartphone. The smartphone acts as an intermediate device between the IoT device and cloud service. Such an architecture is the realization of the *edge computing* concept, where a small intermediate edge device (smartphone) is located on the edge of the Internet network and communicates to the cloud services.

However, the presented design can also act as a standalone solution that can work independently with the IoT device to deliver essential mHealth services. In addition it can use the Internet for synchronization and collaboration purposes. Therefore, it is a realization of the *dew computing* solution.

The presented streaming IoT mHealth solution can be used to build a complete system for continuous real-time remote health monitoring without obstacles for human's daily activities surpassing geographical, temporal, and even organizational barriers.

This idea moves mHealth rapidly as an evolving sector offering great healthcare promise. There still are barriers to unlock the full potentials of such an mHealth solution and mainly depends on the availability of the existing technology, starting from the energy consumption, communication capacity or processing, and storage capabilities.

References

1. Balandin S, Balandina E, Koucheryavy Y, Kramar V, Medvedev O (2013) Main trends in mHealth use scenarios. J Sel Top Nano Electron Comput 1(1):64–70
2. Da Xu L, He W, Li S (2014) Internet of things in industries: a survey. IEEE Trans Industr Inf 10(4):2233–2243
3. Fiordelli M, Diviani N, Schulz PJ (2013) Mapping mHealth research: a decade of evolution. J Med Internet Res 15(5)
4. Free C, Phillips G, Felix L, Galli L, Patel V, Edwards P (2010) The effectiveness of m-health technologies for improving health and health services: a systematic review protocol. BMC Res Notes 3(1):250
5. Gusev M (2017) A dew computing solution for IoT streaming devices. In: 2017 40th international convention on information and communication technology, electronics and microelectronics (MIPRO). IEEE, pp 387–392
6. Gusev M, Dustdar S (2018) Going back to the roots—the evolution of edge computing, an IoT perspective. IEEE Internet Comput 22(2):5–15
7. Gusev M, Stojmenski A, Guseva A (2017) ECGalert: a heart attack alerting system. In: International conference on ICT innovations. Springer, pp 27–36
8. Nastic S, Rausch T, Scekic O, Dustdar S, Gusev M, Koteska B, Kostoska M, Jakimovski B, Ristov S, Prodan R (2017) A serverless real-time data analytics platform for edge computing. IEEE Internet Comput 21(4):64–71
9. Ray PP (2018) An introduction to dew computing: definition, concept and implications. IEEE Access 6:723–737
10. Riley WT, Rivera DE, Atienza AA, Nilsen W, Allison SM, Mermelstein R (2011) Health behavior models in the age of mobile interventions: are our theories up to the task? Transl Behav Med 1(1):53–71

11. Satyanarayanan M (2017) The emergence of edge computing. Computer 50(1):30–39
12. Shi W, Dustdar S (2016) The promise of edge computing. Computer 49(5):78–81
13. Shi W, Cao J, Zhang Q, Li Y, Xu L (2016) Edge computing: vision and challenges. IEEE Internet Things J 3(5):637–646
14. Silva BM, Rodrigues JJ, de la Torre Díez I, López-Coronado M, Saleem K (2015) Mobile-health: a review of current state in 2015. J Biomed Inform 56:265–272
15. Skala K, Davidovic D, Afgan E, Sovic I, Sojat Z (2015) Scalable distributed computing hierarchy: cloud, fog and dew computing. Open J Cloud Comput (OJCC) 2(1):16–24
16. Statista (2018) The statistics portal: mHealth—statistics and facts. https://www.statista.com/topics/2263/mhealth/
17. Wang Y (2015) The initial definition of dew computing. Dew Comput Res
18. Wang Y, Skala K, Rindos A, Gusev M, Yang S, Pan Y (2017) Dew computing and transition of internet computing paradigms. ZTE Commun 15(4)
19. Whittaker R (2012) Issues in mHealth: findings from key informant interviews. J Med Internet Res 14(5)

Chapter 8
A System for Monitoring Water Quality Parameters in Rivers. Challenges and Solutions

Anca Hangan, Lucia Văcariu, Octavian Creţ, Horia Hedeşiu, and Ciprian Bacoţiu

Abstract Automated water quality parameters monitoring systems are essential for state-of-the-art water management, as they enable early detection and fast response to pollution events and provide large amounts of data for decision support systems. In this chapter, we identify and discuss the challenges of implementing a system for monitoring water quality in rivers from continuous data acquisition, to standards compliance and automated pollution detection. Moreover, we describe our complete solution for such a system, implemented on Someş River, that includes data acquisition implemented using WSNs, standard-compliant data storage, data provision services, and automatic assessment of water quality.

Keywords Water quality parameters monitoring · Continuous data acquisition · Wireless sensor networks · Water resources management

8.1 Introduction

One of the main challenges of the twenty-first century is climate change, affecting hydrological cycles around the world in different ways, from catastrophic droughts to widespread flooding [1]. Moreover, the changes in the way land is used, urbanization and demographic growth are creating new challenges in the water management

A. Hangan (✉) · L. Văcariu · O. Creţ · H. Hedeşiu · C. Bacoţiu
Technical University of Cluj-Napoca, Cluj-Napoca, Romania
e-mail: Anca.Hangan@cs.utcluj.ro

L. Văcariu
e-mail: Lucia.Vacariu@cs.utcluj.ro

O. Creţ
e-mail: Octavian.Cret@cs.utcluj.ro

H. Hedeşiu
e-mail: Horia.Hedesiu@emd.utcluj.ro

C. Bacoţiu
e-mail: Ciprian.Bacotiu@insta.utcluj.ro

© Springer Nature Switzerland AG 2021
F. Pop and G. Neagu (eds.), *Big Data Platforms and Applications*,
Computer Communications and Networks,
https://doi.org/10.1007/978-3-030-38836-2_8

181

domain. An interdisciplinary approach is required to meet these challenges, both in the response to critical situations and for long-term planning.

Far-reaching provisions of several European Directives, such as the Water Framework Directive (WFD) [2] and the Environmental Liabilities Directive (ELD), combined with changing attitudes toward environmental protection, are placing a strong emphasis on the acceleration of the development of novel measurement technologies, applications, systems that generate cost-effective, robust and fit-for-purpose information to address both current needs as well as more complex emerging issues such as source determination, ecological health assessment, and pollutants [3]. Sensors and monitoring is a field undergoing rapid evolution facilitated by a wide number of technological advances. This is impacting the status quo of many sectors including the water sector and altering how water is managed throughout the water cycle.

In the governance model of the future, enhanced stakeholder engagement plays a key role in water-related decision-making processes, stimulating active collaboration, public–private partnerships, and increased involvement with water issues [4]. Collaborative decision-making and the inclusion of stakeholder views will lead to improved services and transparency. Awareness-raising measures will have led to well-informed, smart water users who are aware of the value of water and water usage, and stakeholders will be empowered through open access to information. The vision of the European Water-Smart Society is a society in which the true value of water is recognized and realized, and all available water sources are managed in such a way that water scarcity and pollution of groundwater are avoided.

In this context, the development of faster, cheaper, and easier to use pollution detection methods and instruments, e.g., real-time sensor networks enabling comprehensive environmental monitoring and fast response, especially by non-specialists in the field, should be a priority. A vast network of sensors and metering systems for water resources will generate large amounts of valuable data to be used for innovative decision support for governance systems thus enabling advanced water management.

8.2   Water Quality Monitoring Systems Challenges

Continuous real-time water parameters monitoring is a necessity in the context of contemporary water resources management. In the last years, the importance of water monitoring systems has been acknowledged at the highest levels of human society. Many governments are seeking to create national scientific institutions in which consistent and scientifically defensible methods and strategies are coordinated for improving water quality monitoring [5].

Water quality is a measure of the extent to which water resources can be used for different purposes. This measure is based on particular physical, chemical, and biological characteristics. Water monitoring can be done in many ways [6]. Scientists sample the chemical condition of water, sediments, fish tissue, etc., to determine levels of essential water quality constituents such as temperature, pH, turbidity,

dissolved oxygen, nutrients, metals, oils, pesticides, etc. Various biological measurements of the aquatic plants and the forms of animal life in general, as well as the
ability of test organisms to survive in sample water, are also widely used.

Data collected by specialized institutions (state agencies, etc.) are used to build
the assessments that are necessary for making specific pollution control decisions
[7]. Through continuous data collection and processing an automated system can
detect pollution problems, identify the location where pollution control actions must
be focused, and evaluate the progress achieved.

A water quality monitoring system requires the following major components:

- A sensor network that collects data from sensors;
- A data storage component;
- A data processing component that interprets raw data and generates the corresponding information (graphs, statistics, alerts, decision support information,
  etc.);
- A data service that provides raw or processed data to be further analyzed for
  various purposes.

There has been an important research interest toward the development of monitoring systems that use Wireless Sensor Networks (WSNs) [8, 9]. Following this
trend, several research projects use WSNs for monitoring the value of water parameters [10, 11]. WSNs can monitor water quality parameters through the cooperation
of many heterogeneous sensors and represent a cost-effective, more flexible and
environment-friendly solution to automatic continuous monitoring of water parameters. However, using WSNs for water quality monitoring introduces several issues
related to communication, energy consumption, and error detection. Moreover, not
all quality water parameters can be measured using sensors. These values must be
estimated using a distinct approach.

Data collected from sensors is valuable, but to be able to extract information and
provide decision support, sensor data may not be enough. Weather, river geography
data, and a knowledge base regarding water quality may be of great use.

Furthermore, there is the problem of detecting anomalous values that may signal
pollution. Anomaly detection in data sets is a well-established field of research [14,
15]. Several approaches solve the problem of anomaly detection in the context
of sensor networks [16] such as the use of rule-based or nearest neighbor-based
techniques.

An equally important challenge for water parameters monitoring systems is
making the measurements available to people (users) or to other applications that use
the monitoring data. Knowing that there is a huge amount of information that should
be handled as public information, one of the main concerns in the heterogeneous
hydro-world information is to comply with standards.

We will discuss the main challenges in the following sections.

8.2.1 Water Quality Parameters Acquisition Using WSNs

Most applications for monitoring water parameters include a few common elements whose presence is mandatory due to the distinctive features of these applications.

First, the wide geographical area from which data must be collected imposes the usage of WANs (*wide area networks*) of various types. Because of the large variety of such equipment available on the market, so far there are almost no identical solutions reported in the literature. However, there are a few architectural aspects that appear in most applications; a brief survey will be presented hereinafter.

Water monitoring sensor networks can be considered a particular case of ambient sensor networks. Most remote water monitoring WANs have *fixed nodes*, but there are also WANs whose nodes are *mobile* (the nodes are floating on the monitored river, like in [17]).

We will focus on monitoring WANs that have *fixed nodes*. They are usually organized in a tree topology, presenting the following tiers:

- Tier 1: *Sensor nodes*—capture data directly from the water stream or from a geographical region located in the immediate neighborhood of the watershed. Sensor nodes can be *simple* or *intelligent* (some small amount of processing can be done locally on these nodes);
- Tier 2: *Dataloggers* or *computation nodes*—gather data from the sensor nodes, perform some basic computations on them (mainly various filtering), organize them in some pre-established format and transmit them to the upper level in the hierarchy;
- Tier 3: *Community interface nodes*—as the water monitoring WAN usually aims at fulfilling a function that is of general interest for the human community (early flood warnings, accidental or permanent pollution alerts, etc.) a crucial aspect is to make this information available to a large public. This can be done through various internet-based services (RSS, web portals, etc.), SMS messaging, paging, etc.

As shown in Fig. 8.1, *redundancy* is an important feature (as most researchers agree) that must be implemented when using WANs. Also, there are two types of communication involved:

1. *Local communication* between sensor nodes and dataloggers. This kind of communication is implemented for short-range (from hundreds of meters to a few kilometers) data transmission [9, 10], and it occurs with a relatively high frequency (for instance every 15 min);
2. *Long-distance communication* between dataloggers and community interface nodes. This communication occurs more rarely (only when a significant amount of data is gathered—for instance, every one or two hours).

Of course, the radio frequencies used for the two types of communications must be different (900 MHz for *local*, 144 MHz for *long-distance communication*, in [18, 19].

The transmission equipment usually includes radio towers of a few meters height (5 m in [18, 19]). It is also necessary to take very serious equipment protection

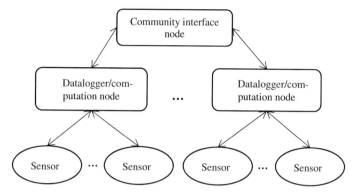

Fig. 8.1 Generic architecture of a water monitoring wireless sensor network

actions (including camouflage, usage of PVC pipes to protect the cables and of metal boxes for securing the sensors, various anti-theft methods, etc.) as many cases were registered when people stole or damaged the outdoor sensors or antennae.

Another important issue is *energy consumption*. All outdoor equipment must be powered by solar panels or very long-life batteries. Solar panels are more vulnerable to hooligan attacks and to weather conditions but can be the ideal solution for some geographic regions. In both cases, special hardware is needed to monitor the battery level and human maintenance operations are mandatory.

A very important part of water parameters monitoring is constituted by *high-frequency water quality monitoring and estimation*. It is impossible to sample all the potential pollutants in a watershed—in fact, as mentioned in [20], for many water quality constituents (e.g., sediment and phosphorus), sensor technology does not currently exist for making high-frequency measurements of concentrations in-situ. An alternative solution consists of monitoring only a small set of parameters called *surrogates*. Then, using some mathematical relationship, the measured values obtained from the *surrogate sensors* are converted into estimates of the variables of interest (see [20–24] for a detailed description).

The "standard" (most commonly used) set of parameters that are of interest is composed of: *water temperature, discharge, specific conductance, turbidity, pH*, and *dissolved oxygen*. Through adequate relationships deduced by conducting laboratory experiments in parallel with sensor-based measurements over large periods of time, the high frequency monitoring of these parameters can provide information about many water quality constituents. Such relationships are inferred using specific mathematical methods like least squares regression or other methods within statistical software [20].

Normally, measurement errors are mentioned in the sensor's description from the manufacturer's catalogue, but they can also be quantified using specific instrument tests. Measurement errors can be quantified by taking replicate samples and making new measurements to derive the components of the measurement error variance [25]. To measure the error in regressions one can use some specialized methods such as the coefficient of determination (R2), the root mean square error (RMSE or MSE), and the relative percent difference (RPD), which provide an indication of the uncertainty in the estimates [23, 24, 26].

Many papers have been published in this field and to the best of our knowledge the main guidelines presented in this section were followed by many of the research teams. Among them, some of the most serious and complete projects could be considered the Early Warning Flood Detection System developed by Basha, Rus, and Savela from MIT [18, 19], which includes all the elements of a complete water monitoring system but also uses prediction models and algorithms for anticipating the sensor values; the Little Bear River sensor network developed by Horsburgh et al. [20], which uses surrogates sensors; the project developed by *The Susquehanna River Basin Commission* (SRBC) from Harrisburg, Pennsylvania [27], which covers an area in which intense hydrofracking activities take place.

8.2.2 Pollution Detection

The detection of abnormal events in environmental monitoring is based on analyzing the values obtained from sensors and the correlations between these values. It is important to differentiate between anomalies caused by faulty sensors that provide erroneous values and anomalies caused by events.

In the special case of pollution detection in rivers, one must consider certain correlations between water quality parameters. Moreover, water quality parameters may be correlated with the weather, seasons or river geology. For example, high turbidity is usually detected during and after a rainfall and it causes an increase in temperature and a decrease in dissolved oxygen. This will cause damage to the flora and fauna of the river. Conductivity and pH levels are specific to each water stream due to the soil and geology. Therefore, the change in pH and increased conductivity levels signal the presence of polluting chemicals such as nitrate, phosphate or sodium. Water temperature accepted value interval varies with the seasons. We can conclude that pollution detection in rivers is not trivial because it involves contextual anomalies [16] that can be detected only by first evaluating the context (the condition of the anomaly).

Three types of correlations must be considered in data sets that are collected from sensors:

- Time correlations—between values measured by the same sensor;
- Space correlations—between values measured for the same water quality parameter, but in different locations;
- Functional correlations—between values measured for different water quality parameters, in the same location.

The detection of an outlier value based on the admitted ranges for each water quality parameter may not determine that there is an event, until the detection of time, space, and functional correlations. If the specific correlations exist, there can be a conclusion that an event appeared.

To be able to take advantage of time and spatial correlations, the measurements acquired from sensors have to be made continuously in subsequent locations on the river shore for each of these parameters. To be able to take advantage of functional correlations, in each location, several water quality parameters that are connected should be measured.

To differentiate between erroneous measurements and actual events, the correlation of the values measured at subsequent locations has to be used. If an event appears at one location, then the measurements downstream for the same parameter will be correlated. Moreover, values showing events are time-correlated. In case of errors, there are no spatial correlations. A faulty sensor can give unpredictable readings. Another solution to error detection is to use redundant sensors and to check the correlations between values measured at the same time and in the same location for the same property.

8.2.3 Standards for Hydrographic and Monitoring Data

There are significant efforts made worldwide toward establishing standards for water quality monitoring. In the European Union, the Water Framework Directive (WFD) establishes, among others, a guide for monitoring the quality elements of rivers, to assure the interoperability between different platforms [2]. The guide presents the appropriate selection of quality elements and parameters for rivers, lakes, transitional waters, and coastal waters to support the implementation of the WFD and additional recommended quality elements, which have been identified by Member States for that particular waterbody type. Due to interoperability requirements, new platforms must be compliant with the INSPIRE Directive of the European Union [12] and with the OGC Sensor Observation Service Standard (SOS) [13].

Since we monitor water quality in rivers, to provide information about the water quality parameters in specific locations, we need to represent the geography of the river. We also need to store sensor observations for monitored locations on the river. Moreover, we have to be able to make our data available in standard format as required by EU Directives.

EU INSPIRE (INfrastructure for Spatial InfoRmation in Europe) Directive was initiated to create a European SDI. The first objective of INSPIRE is to provide better common environmental policies in Europe. To be able to provide compatibility between spatial data from EU states, a set of implementation rules were adopted. These Implementation Rules refer to:

- Metadata;
- Data structures specifications;
- Network services;
- Data and services sharing;

- Monitoring and reporting.

Furthermore, they specify that the implementations should be based on OGC and ISO standards.

To be able to provide interoperable components in the proposed system, we comply with INSPIRE Implementation Rules, as follows:

We use INSPIRE Hydrography model for the representation of Somes River;

We use OGC Sensor Observation Service (SOS) to maintain observations received from sensors.

The INSPIRE Hydrography model is a standard data model for representing physical water elements for the purpose of creating maps, models, as support for analysis and for management and reporting in the context of WFD. There are three available schemas:

- Hydro—Physical Waters (for maps);
- Hydro—Network (for models and analysis);
- Hydro—Reporting (for management and reporting).

The OGC SOS Standard provides specifications for a web service interface that allows querying for sensor observations (measured values) and provides the means for registering sensors and for recording the sensors' readings. The standard requests accepted by the SOS [8] component that we have implemented are the following: query observation, insert observation, register sensor, remove sensor.

Data measurements from sensors are stored in an SOS Server that provides a standard data structure and database. Some implementations use a Postgre database with a PostGIS extension and others use Oracle and Oracle Spatial extension. For our system, we use Postgre with PostGIS extension. The interaction with the database is done through web services, as specified by the SOS standard.

The standard SOS data model includes the following concepts:

- Feature of interest;
- Procedure (e.g., measurement procedure);
- Observation;
- Phenomenon;
- Composite phenomenon;
- Quality (it refers to the observation);
- Offering.

Any monitoring system that uses sensors should be mapped on this standard data model. This way, any application that includes an SOS client can inquire about sensor measurements.

8.3 A Service-Based System Architecture for Water Quality Monitoring

Having identified and analyzed the most important issues for the water quality monitoring domain, we are able to point out the most important requirements of the main components of a water quality monitoring system.

Data acquisition from multiple sources, using multiple acquisition methods insures the extraction of meaningful information about water quality and provides better support for the detection of events such as pollution or floods. Measurements made continuously by sensors installed along the river are the most reliable and large source of data. However, not all parameters that are used for water quality assessment can be measured by sensors. Those parameters have to be estimated through computation and added to the database (e.g., river discharge). The process of water parameter estimation often relies on some measured values (by sensors or human observers) as well as on hydrographic data (e.g., river geography) and contextual data (e.g., weather data, river vicinity geography). Moreover, for the interpretation of recently acquired water quality data, historic data could be of great importance as a source for pattern extraction. Historic measurements, as well as hydrographic and contextual data, are usually provided by third parties. Problems of availability, reliability, and non-standard format may arise.

The measurement of water parameters by sensors installed along the river insures continuous online acquisition of values that will be used for water quality assessment. Data should be available online as soon as measured by the sensor. Wireless sensor networks are one of the most cost-effective and easy to install infrastructures that can be used for this purpose. The same parameters should be measured in several subsequent locations along the river to be able to detect space correlations between measurements made for water parameters. Time and space correlations between measurements are of great importance for the detection of events (e.g., pollution) as well as of erroneous measurements.

The sensors provide measured values for water parameters continuously, producing a large amount of data that must be stored in a reliable manner. It is recommended that the format and the access method to data are standard. Since these kinds of data are hard to obtain, this will insure interoperability with other applications that use water quality measurements. Moreover, standardization insures the reusability of data processing and of data provision services. Data provision services extract values from the storage according to a preset filter and present it to the user in a standard format. Data processing services estimate the value of water quality parameters that cannot be measured by sensors but can be computed using equations that combine multiple values measured by sensors and sometimes values measured manually by humans.

Information services use data acquired from different sources, raw or preprocessed, and expose it to users in a meaningful way in multiple formats:

- Up to date measurements for individual water parameters in different locations depicted on a map;

- Charts showing the evolution of measurements for individual water parameters;
- Statistics, making use of historical values.

However, many users would not be able to interpret water parameters values and would not be able to recognize the correlations between parameters to assess the quality of the water. The quality of information presented to the users would greatly benefit from an automated expert system able to assess water quality and even give pollution alerts or estimate the concentration of pollutant over time.

To support our view of a water quality monitoring system, we propose the service-based architecture depicted in Fig. 8.2.

The proposed system architecture has three layers that organize the components into: data sources, data processing, storage and data provision services, and information services. Data is collected from multiple sources (e.g., sensors, manual measurements and third-party data providers). Raw (unprocessed) data can be accessed through SOS standard services or the data filtering services. Information extracted from collected data is provided through information services.

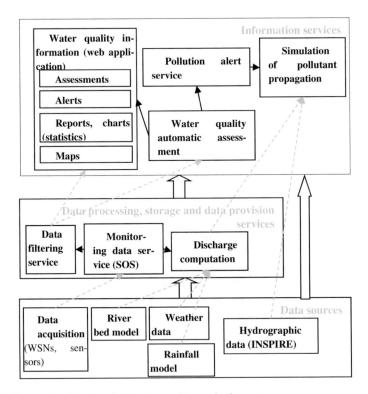

Fig. 8.2 Proposed architecture for a water quality monitoring system

8.3.1 Data Sources

The proposed architecture includes several data sources that will be used to extract information and for automatic water quality assessment:

- WSNs installed in different locations on the river;
- Hydrographic data;
- Weather data;
- River bed measurements;
- Rainfall model data.

The main sources of data are the WSNs installed on the river shore that provide continuous measurements. Water parameters are measured by sensors submerged in the river. The focus of the data acquisition sub-system is to provide values for the most common water parameters that can be used for the evaluation of water quality:

- Water temperature;
- pH;
- Conductivity;
- Dissolved oxygen;
- Turbidity.

On the river shore, there is need for a local data logger that gathers data from all sensors in the area. WSNs solve the problem of logging data from sensor nodes that have several sensors attached to measure different water parameters. The WSN gateway retransmits collected data to an application that stores it. A feasible solution for handling long-distance communication from the river shore is GSM communication.

Sensor nodes, however, can be spread on an area of more or less 1 km^2 to be serviced by a single gateway. But to be able to take advantage of spatial correlations between measurements for the same parameter, sensors of the same type must be installed several kilometers away from each other. In this case, there will be more than one local data acquisition sub-systems on the river shore, each with its own gateway.

Another source of data is a service or a database that provides hydrographic data. By hydrographic data, we understand: the river identifier, its geographical name, the river topology represented as a network with nodes and watercourse links, its localization, its flow direction, and its length. Usually, hydrographic data are provided by specialized national institutions. However, in our case, we constructed our own hydrographic model for a specific river. Weather data is also obtained from specialized services.

Another category of data is represented by parameters that rarely change. These are obtained through measurements made manually or estimated using models (e.g., river bed, rainfall).

8.3.2 Data Storage, Processing and Data Provision Services

One of the important requirements of this architecture is interoperability. The first step toward meeting this requirement is to comply with standards for data formats and data provision.

Because Romania is a member of the European Union, we have to comply with the INSPIRE Directive for keeping hydrographic data. The INSPIRE Directive provides data model specifications for spatial data, including hydrographic data. We use a sub-set of the Hydrography Model provided by INSPIRE, mainly the Hydro-Base and Hydro-Network Application Schemas [12] to store data about the monitored river basin.

To manage sensor data in an interoperable way, we use OGC Sensor Observation System Standard (SOS). The SOS Standard provides specifications for a web service interface that allows querying for sensor observations (measured values) and provides the means for registering sensors and for recording the sensors' readings [13]. Since the WSNs do not use implicitly the SOS format, raw data received from sensors is translated in SOS format. An SOS client is then used to register the sensors and to store the sensor observations on a private installation of the Sensor Observation Service. The SOS exposes a set of predefined data services that will provide recorded data according to standard specifications. However, since usually there is need for filtered data, the architecture contains a special data filtering service that can be queried by clients that need to access values measured by the monitoring system. The data filtering service receives a set of parameters (e.g., measured parameter, GPS coordinates, value ranges, time ranges) based on which data is extracted from the SOS server.

Data processing services have the role of computing and providing values that are not measured by sensors. Such a service is the discharge computation component. The discharge is an important parameter used to estimate pollutant downstream propagation and floods. The computation can include an estimation of additional water quantities derived with a rainfall model based on weather data.

8.3.3 Information Services

The practical purpose of the water quality monitoring system is to provide public information about the water quality, alert the public and authorities when critical events (e.g., pollution) occur and, if possible, suggest some actions that can be taken to manage critical situations. To address these objectives, the proposed architecture integrates:

- A component for the automatic assessment of water quality;
- A pollution alert service;
- A simulation component for pollutant propagation estimation.

Finally, a web application presents the information to the average user. Inside the application, the user can see charts and reports showing the evolution of water quality parameters over time and current sensor readings shown on maps. The user can sign up for pollution alerts and can view the results of automatic assessments over water quality. If a pollution event occurs, the user can use the simulation to view the estimation of pollution propagation.

The component for the automatic assessment of water quality is in fact a rule-based decision system that has two steps:

1. Retrieves sensor observations from the SOS server and labels them based on sets of rules specific for each parameter. This step has the role of detecting anomalies in time series.
2. Labeled values are processed according to another set of expert rules that have the objective to find correlations between different parameters values which can signal critical situations. The second step will be able to emit suggestions for actions if a critical situation is detected.

The pollution alert service sends alerts to subscribed users if a pollution situation is detected. The user will receive a message containing information about the type of pollution that occurred. The web application will show the current alert (if any) and the history of previous alerts.

The simulation component for pollutant propagation estimation computes the pollutant concentration in several consecutive locations on the river starting with the point in which pollution has been detected. The concentration is computed based on the value detected by sensors, pollution propagation model computed for a specific river and pollutant. Pollution propagation depends greatly on the discharge and river physical parameters (e.g., width, depth, and slope). Hence, the propagation model parameters are specific to each river.

To sum up, the proposed water quality monitoring architecture offers support for the following features:

- Continuous monitoring made with WSNs;
- Data collection from multiple data sources to improve information extraction;
- Measurements provided online as soon as they are collected;
- Data storage and presentation through services in standard formats established by the European Union;
- Estimation of parameters that cannot be obtained by direct measurements;
- Simulation of pollutant propagation;
- Automatic assessment of water quality based on rules;
- Information provision through a web application and web services.

8.4 A Pollution Detection System for Somes River

The pollution detection system for Somes river is built on the proposed water quality monitoring architecture. The features supported by the architecture are adapted to the specific purpose of pollution detection in the Somes river. The role of this system is to monitor the water quality parameters of Somes river with the help of a wireless sensor network (WSN) installed on the river shore and to issue alerts if the monitored parameters are out of range of the accepted values range.

The monitored parameters are the following:

- Water temperature;
- Conductivity;
- Turbidity;
- Dissolved oxygen;
- pH;
- Water pressure (needed for computation of discharge).

The pollution detection system for Somes river has the following components:

- Data acquisition, done using a WSN;
- SOS server that stores monitoring data in a standard format;
- Hydrographic data that stored in an INSPIRE compatible format;
- Discharge computation for a segment of Somes river;
- Simulation of pollutant propagation adapted to Somes river;
- The data filtering service extracts data from the SOS server according to user-requested filters;
- The rule-based automatic assessment of water quality;
- The alert service issues alerts, if there are parameters that do not fall within the allowed value ranges;
- The water quality information web application presents the data recorded by the WSN through tables and charts (data can be filtered by the user), shows on the map the location of the points where the monitoring is done and alerts users of pollution incidents.

8.4.1 Data Acquisitions and Storage

The monitoring system is based on a sensor acquisition system for measuring parameters of interest, local processing, data transmission, and storage. Parameters that are not measured using sensors are estimated based on surrogate parameters. There is also the possibility of retrieving data from external sources. Measurements were made at the river's point of entrance in the city of Cluj-Napoca.

The acquisition and transmission of measured sensor data are performed within our system using a WSN provided by National Instruments [28]. Figure 8.3 presents the structure of the monitoring system.

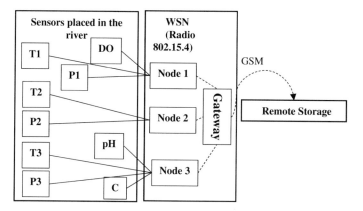

Fig. 8.3 Data acquisition through the wireless sensor network

The system includes sensors that are specific for measuring the pressure (P), temperature (T), conductivity (C), pH, and dissolved oxygen (DO). All the sensors are linked to the nodes of the WSN. A gateway takes over the data from the nodes and transmits the information to a server where the database is stored.

LabVIEW WSN has a simple API to perform the data exchanges between the node and the host system. Support for I/O values transmission, messages made from characters series and node's information are also provided. All data reception is made with the Radio Message item. Each received message is stored in a FIFO memory buffer existing on the gateway, which can store up to 40 messages. Then the gateway tries to deliver every message to the corresponding node. If the FIFO memory is full, then an error message will be generated.

The sensor network periodically samples sensor measurements and performs the data transmission to the base station (gateway). The sensor nodes perform different operations before sending the data to the gateway. The first step is *Sampling*, which implies reading the values from the sensor at predefined time intervals. The measuring process of the sensor values implies converting the parameters of a physical phenomenon in electric current specific values (i.e., voltage and/or amperage), according to the value of the external stimulus. The next step is *Signal conditioning*. First, this has the role of transforming the values gathered from the sensor itself at time intervals permitting the use by other digital systems. Most of the times, a signal amplification is performed, but it is also possible to apply band-pass filters according to the needs (for instance, for eliminating noises of a given frequency, etc.).

After conditioning the signal is converted into a digital one; this process is called *Signal conversion* and it is done using an analog–digital converter (ADC). The data obtained after the conversion can be further used to compute the signal's average, standard deviation, minimal and maximal values, etc.

The *Transmission* toward the gateway is the last step that is performed on the nodes. This operation is a huge energy consumer and thus the lifetime of the nodes is significantly reduced, proportionally to the number of transmissions that are done.

Therefore, the transmission module is normally disabled. It is enabled or activated only during this process.

The gateway is the component that gathers data from all the nodes and controls them. It must keep two active connections: one with the sensor nodes, using a short-range protocol (<1.6 km), and one with a data server to which it will send the gathered data. The connection with the data server is permanent, unlike the other ones (with the sensor nodes) which send data at larger time intervals, then go idle. The gateway is remotely programmable and can be reconfigured in case functionality corrections are desired.

At each measuring performed on the nodes, the gateway saves the received data in a FIFO structure, storing the last n measurements. The storing limit is done by the maximal space available in the gateway's memory. The data on the gateway are made available in two formats, XML (eXtensible Markup Language) and CSV (Comma Separated Values), in corresponding files that are generated from the measuring queue (the above described FIFO). The data coming from the gateway are available on a web server and can be accessed via the HyperText Transfer Protocol (HTTP). Data are then registered to the SOS server using an SOS client.

The installation of the data acquisition system has been done on Somes river and actual measurements were made in 3 locations as follows:

- Three measuring points were chosen in an area at the entrance in Cluj-Napoca, an area that did not have hydrographic works and where there were no meanders;
- The sensor immersion points were set at a distance of 100 m between them and 8 m of the river's shore;
- Topographical measurements have been made to determine the profile of the river bed in the area of the measuring points;
- The sensors have been installed in water by using metal poles as presented in Fig. 8.4 (on the right side);

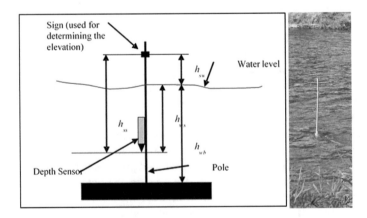

Fig. 8.4 Measurement assembly for pressure (design on the left and installed on the right)

- A wireless node with connected (wired) sensors was installed at each measuring point;
- A gateway was used to communicate collect the measurements from the 3 nodes;
- The gateway was connected to the server using GSM communication.

8.4.2 Discharge Computation

River discharge computation was made using Manning's equation (8.1) [29]. The parameters were determined manually (topographical measurements) and by sensor measurements (pressure sensor).

Having three measurement points we obtain two values for the slope. Using these values, we compute the average of the slope used to determine the discharge according to (8.1).

$$Q = \frac{1}{n} A R_h^{\frac{2}{3}} S^{\frac{1}{2}} 1 \qquad (8.1)$$

where:

Q = the river discharge;

A = the river section area at the measurement point;

R_h = the hydraulic radius defined like section area per wetted perimeter;

S = the slope of the hydraulic grade line;

n = the Manning coefficient (depends on the river bed characteristics and is river specific) [30].

Figure 8.4 shows (on the left side) the measurement assembly for the pressure sensor used in all three points. The elements that can be observed are:

- The vertical pole, with a hose clamp in the upper part used to aim at slope determining;
- Pressure sensor (depth), on the vertical pole, under the river water level.

The elements that can be observed are:

- The vertical pole, with a hose clamp in the upper part used to aim at slope determining;
- Pressure sensor (depth), on the vertical pole, under the river water level.

Distances indicated have the following meanings:

- $h_{ss} = h_{sign\text{-}sensor}$—the distance from the hose clamp to the pressure sensor;
- $h_{sw} = h_{sign\text{-}water}$—the distance from the hose clamp to the river water level;
- $h_{ws} = h_{water\text{-}sensor}$—the distance from the river water level to the pressure sensor;
- $h_{wb} = h_{water\text{-}bed}$—the distance between the river water level and the bed of the river (practically, the depth of the water at that point).

Table 8.1 Measured and calculated parameters for discharge

Segment	S1–S2	S2–S3
Segment length	105.277 m	86.046 m
Segment misalignment	0.405 m	0.09 m
Segment slope	0.0038	0.0010
River section area	18.28 m^2	
Wetted perimeter	34.976 m	
Hydraulic radius	0.522 m	
Total distance between sensors	191.323 m	
Total slope	0.495 m	
Average slope	0.00258	
n coefficient value on Somes river	0.04	
Discharge	**15.08 m^3/s**	

The topographical measurements were used to determine the river section area in the measurement points. For the three measurement points, in the order S1-S2-S3 (the central point is S2), the determined values are:

S1: $h_{ss} = 53$ cm, $h_{sw} = 33$ cm, $h_{ws} = 20$ cm, $h_{wb} = 66$ cm.
Absolute Z (against sea level) = 348.65 m.
S2: $h_{ss} = 80$ cm, $h_{sw} = 60.5$ cm, $h_{ws} = 19.5$ cm, $h_{wb} = 36$ cm.
Absolute Z (against sea level) = 348.59 m.
S3: $h_{ss} = 74.5$ cm, $h_{sw} = 37.5$ cm, $h_{ws} = 37$ cm, $h_{wb} = 63$ cm.
Absolute Z (against sea level) = 348.60 m.

Table 8.1 presents the measurements and the computed parameters for the Somes River upstream of the town of Cluj-Napoca and the value of the discharge computed using Manning's formula (8.1).

The river discharge is stored in the database to be used for pollutant propagation simulation component and in the process of decision-making for automatic assessment of water quality.

8.4.3 The Rule-Based Automatic Assessment of Water Quality and Pollution Alert Service

The pollution alert service will emit alerts if an event is detected based on the monitored parameter values. Our approach for detecting events while monitoring water quality parameters is a two-step rule-based system. In the first step, the parameter values are labeled based on a set of rules that take into consideration time and space correlations between the values measured for each parameter. In the second step, a second rule-based component assesses the functional correlations between several parameters to detect events such as river shore erosion, floods or chemical pollution.

The first step of the rule-based event detection system is focused on the detection of anomalies in the time series of each measured parameter, at each location. These anomalies are values that are outside the accepted value interval, which may signal an event. Labeling rules are different for each parameter, not only because accepted value intervals and correlation rules differ, but also because some parameters' accepted value intervals are variable based on the context in which they are measured (e.g., normal values for water temperature vary based on season). The labels assigned to the measured or computed values place them in one of the following categories: normal, low, high, very high.

Labeled values are passed to the second step rule-based component that will be able to detect actual events based on correlations between several parameters values. For example, river shore erosion may be detected based on high turbidity and high discharge. River shore erosion may signal the risk of floods. If the river shore is near an agricultural land, in the presence of a flood, there is a high risk of nitrate and nitrite pollution. High turbidity is usually detected during and after rainfall and it causes an increase in temperature and a decrease of dissolved oxygen. This will cause damage to the flora and fauna of the river. Conductivity and pH levels are specific to each water stream due to the soil and geology. Therefore, the change in pH and increased conductivity levels signal the presence of polluting chemicals such as nitrate, phosphate or sodium.

By applying similar water quality assessment rules the second step component will be able to identify various types of events. The performance of event detection is heavily influenced by the quality of preliminary value labeling. An increased spectrum of categories (label types) should improve the assessment process.

The component in charge of the automatic assessment of water quality is a distributed multi-agent system [31] as presented in Fig. 8.5. The information flow inside the system starts with the data agent that interrogates the SOS server in order to get the newest measurements received from the WSN, and then it transmits the converted data to the inference engine, to the decision agent, and to the GUI. The next step is done by the inference engine, which transmits to the decision agent and the GUI the results of the activation of the rules; then, the decision agent sends the values of the decisions to the GUI [32].

The user can send commands via the GUI, which arrive at the configuration agent, which in turn sends commands toward the rest of the agents. Besides all this, the configuration agent autonomously sends configuration commands to the other agents on certain particular events that require a change of functionality at the level of one or more agents.

For inter–agent communication we are using a message-passing approach. The distributed agents communicate through messages that can either be transmitted directly, from sender to receiver, or in case there is no direct link between those two, via other agents that are forwarding the message until it reaches its destination.

Agents will form client–server connections between them, using TCP sockets. In our system, an agent can be a server and a client at the same time.

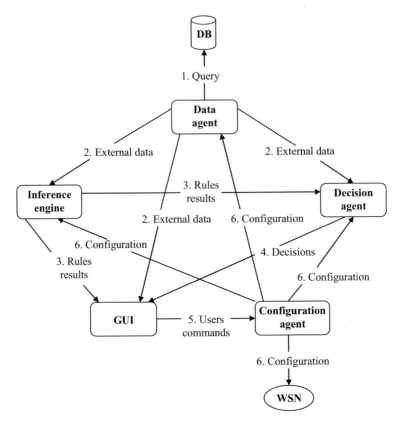

Fig. 8.5 Decision system component architecture and information flow

As for the agents' communication language we have chosen the FIPA ACL standard because it has rigorously defined formal semantics and uses message exchange between agents [33].

The data agent has the role of converting the data extracted from the database and forwarding them to the agents according to their needs. It also has the role of inserting in the database (in the decision table), the data received from the decisions agent, which constitute information about events, scenarios or observations related to the specific river basin.

The inference engine has the role of evaluating the performed measurements in order to detect the values associated to water quality in the form of a scenario. The scenario includes different elements like: exceeding the threshold, details about the current status of the sensors and the occurrence of some meteorological phenomena like pollution with various substances, floods, etc.

The inference engine is composed of the three main components of a rule-based expert system, namely: the working memory, the rules, and the inference engine

Fig. 8.6 Rules subsystem

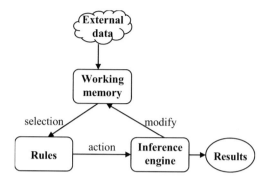

itself. The rules engine is a reactive agent that acts like a reflex, since it responds to stimuli with previously known values, without keeping an internal state (Fig. 8.6.).

The decisions agent is the most important because it contains knowledge about solving some problems and fighting some phenomena (e.g., pollution situations) that affect the quality of the monitored water resources. The agent is the one that offers the manager, via the GUI, solutions for solving any problem that can occur in the context of water quality management. Although it depends on the rules engine (in order to be able to analyze water quality and to make decisions upon the scenario detected by it), it can, thanks to the data received from the data agent, emit warnings for the manager about the exceeding of some specific values that represent safety thresholds for the measured parameters.

The decisions agent is a pro-active agent which tries to reach a long-term goal. For that it looks for the adequate action for each situation by trying to predict the effects of that action and by improving its knowledge after each action performed—it reflects on the effects of the previous action.

The decisions agent detects the dangerous scenarios and analyzes their gravity in order to propose actions that will fight as effectively as possible the negative effects of the detected event. The decisions agent detects scenarios like: floods, pollutions by various substances, high/low levels of the river discharge, turbidity, electric conductance, pH, dissolved oxygen, and other measured or computed parameters.

The configuration agent has the role of configuring various functionality settings of the decisions agent, of the rules engine, and of the data agent. This agent can receive configuration commands from the final user by means of the graphical interface. The configuration agent is a reactive one; it receives information from the rest of the agents and modules, then it reacts by sending configuration messages toward the agents that require changes in order to function correctly or to improve their efficiency.

The graphical user interface (GUI) is the only component that the final user can directly access, so it constitutes his/her communication point with the decision support system. The GUI has the role of presenting the user all the data gathered from the decision support system, and especially data received from the decisions agent, the data agent, and the rules engine.

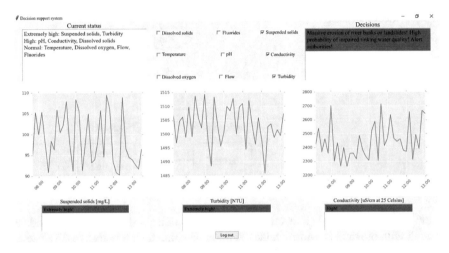

Fig. 8.7 Pollution due to river shore erosion critical alert

Figure 8.7 shows an example of automatic assessment made based on very high values for turbidity and suspended solids and high values of pH, conductivity, and dissolved solids. The rule-based decision system concludes to emit an alert for pollution due to river shore erosion.

8.4.4   Simulation of Pollutant Propagation

If a pollution event is detected, the simulation of pollutant propagation can estimate the concentration of pollutant downstream from the measurement point, over time. The simulation component has to be provided the location pollution was detected, initial pollutant concentration, and a time span (minutes) for the simulation step. The simulation runs several consecutive steps to obtain the estimation of pollution distance and concentration. Figure 8.8 shows the results of the simulation. We used

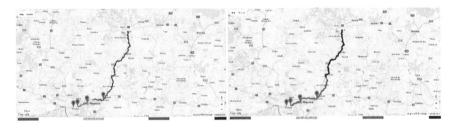

Fig. 8.8 The pollutant propagation simulation first step (left) and result (right)

a color code to highlight the pollutant concentration values on different segments of the river, on the map.

The propagation model used was derived for Somes River by [34]. Since the hydrographic model used in this case was not a standard one and furthermore, it was not INSPIRE compatible, we provided a transform method between the models to be able to use INSPIRE compatible data as input and to provide a compatible output as well. The propagation model uses a specification of the river in which points are described by their distance from the river's spring coordinates. INSPIRE describes the river as a network of nodes and edges located through latitude and longitude parameters. For a point located on the river, we transform latitude and longitude to distance, and backwards. For this, we use the Haversine equation [35].

Discharge values needed for pollutant propagation estimation are extracted from the database. The discharge is computed by the dedicated component presented in Sect. 8.4.3.

The hydrographic model is specific for a certain segment of Somes River and chloride pollution. To be able to provide similar simulations for other rivers and other types of pollutants, the propagation model has to be changed. Since we have developed the simulation as a distinct sub-module that receives input and provides output to a standard environment, the system architecture will not require essential changes in the case in which we decide to implement it on a different river or for other types of pollutant.

8.4.5   The Water Quality Information Web Application

By accessing the water quality information web application, a user can view:

- The measurement points location on the map;
- Recent measurements collected by the WSN;
- Extract and view filtered data from the SOS server;
- Pollution alert current status and history.

The user can also subscribe for email pollution alerts.

The main page of the web application (see Fig. 8.9) shows on the map the measurement locations correlated with the parameters that are measured at each location, the latest measurements, a data filter, and latest pollution alerts.

The user can fill in a form to extract and view filtered data from the SOS server. The filter options are:

- Measured property;
- GPS coordinates of the measurement location;
- Measurement ranges;
- Timestamp ranges.

The filter options are automatically passed to the data filtering service. The results are shown in a table as well as in a graphic chart format as can be seen in Fig. 8.10.

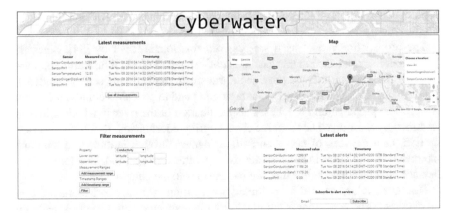

Fig. 8.9 Main page of the water quality information web application

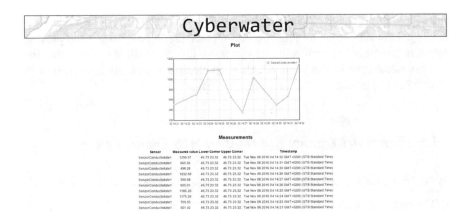

Fig. 8.10 View recent conductivity measurements and chart

8.5 Conclusions

Water quality monitoring systems are of great importance for an improved management of water resources in the context of dealing with climate change. To be able to sustain a vast network of sensors and metering systems, the development of faster, cheaper, and easier to use methods and instruments for data acquisition, processing and information extraction is required.

In this chapter, we describe our approach for the development of water quality monitoring systems in rivers. We start by identifying and discussing the most important challenges when dealing with water quality monitoring. Having identified the issues, we present our vision for water quality monitoring and we point out the requirements of the main components of a water quality monitoring system focused on pollution detection. Moreover, we propose an architecture that is able to support the most important features of a water quality monitoring system. Finally, we present the development of the main components of a pollution detection system customized for Somes river. We describe a complete solution that includes a data acquisition sub-system implemented using WSNs, data storage solutions that are standard-compliant, data provision services, automatic assessment of water quality and pollutant propagation simulation.

References

1. Chemistry and water: challenges and solutions in a changing world. In: A white paper from the 6th chemical sciences and society symposium (CS3), Leipzig, Germany, September 2015
2. Water framework directive common implementation strategy, guidance on monitoring for the water framework directive, Final Version, 2003
3. WssTP–Sensors & Monitoring Report. State of the art and research needs, June 2012
4. WssTP Water Vision–The value of water, WssTP Water Vision 2030, WssTP, Brussels, 2017, ISBN: 9789028362130
5. McDonnell R (2008) Challenges for integrated water resources management: how do we provide the knowledge to support truly integrated thinking? Int J Water Resour Dev 24(1):131–143
6. EPA, United State Environmental Protection Agency. http://water.epa.gov/type/watersheds/monitoring/monintr.cfm
7. Ciolofan S, Mocanu M, Ionita A (2013) Cyberinfrastructure architecture to support decision taking in natural resources management. In: Proceedings of the 19th international conference on control systems and computer science, CSCS19, 1st international workshop on cyberinfrastructures for natural resources management, CyRM-2013, May 29–31, 2013, Bucharest, Romania, pp 617–623
8. Porter JH et al (2009) New eyes on the world: advanced sensors for ecology. Bioscience 59(5):385–397
9. Quansah JE, Engel B, Rochon GL (2010) Early warning systems: a review. J Terr Obs 2(2):22–44
10. Jiang P, Xia H, He Z, Wang Z (2009) Design of a water environment monitoring system based on wireless sensor networks. Sensors 9:6411–6434
11. Gunatilaka A, Moscetta P, Sanfilippo L (2007) Recent advancements in water quality monitoring: the use of miniaturized sensors and analytical measuring techniques for in-situ and on-line real time measurements of surface water bodies. In: Proceedings of the international workshop on monitoring and sensor for water pollution control, Beijing
12. INSPIRE (2015) http://inspire.ec.europa.eu/index.cfm
13. Sensor Observation Service (2015) http://www.opengeospatial.org/standards/sos
14. Zhang Y, Meratnia N, Havinga P (2010) Outlier detection techniques for wire-less sensor networks: a survey. IEEE Commun Surv Tutorials 12(2):159–170
15. Rassam MA, Zainal A, Maarof MA (2013) Advancements of data anomaly detection research in wireless sensor networks: a survey and open issues. Sensors 13(8):10087–10122
16. Chandola V, Banerjee A, Kumar V (2009) Anomaly detection: a survey. ACM Comput Surv 41(3), Article 15
17. Floating Sensor Network, http://float.berkeley.edu/
18. Basha E, Rus D (2007) Design of early warning flood detection systems for developing countries. In: Proceedings of the conference on informations and communication technologies and development, Bangalore, India, December 2007, pp 1–10
19. Basha E, Ravela S, Rus D (2008) Model-based monitoring for early warning flood detection. In: Proceedings of the 6th ACM conference on embedded networked sensor systems (SenSys), Raleigh, NC, November 2008, pp 295–308

20. Horsburgh JS, Spackman JA, Stevens DK, Tarboton DG, Mesner NO (2010) A sensor network for high frequency estimation of water quality constituent fluxes using surrogates. Environ Model Softw 25(9):1031–1044
21. Christensen VG (2001) Characterization of surface-water quality based on realtime monitoring and regression analysis. Quivira National Wildlife Refuge, South-central Kansas, December 1998 through June 2001. USGS Water Resources Investigations Report 01-4248, 28 pp. http://ks.water.usgs.gov/pubs/reports/wrir.01-4248.pdf
22. Christensen VG, Jian X, Ziegler AC (2000) Regression analysis and real-time water-quality monitoring to estimate constituent concentrations, loads, and yields in the little Arkansas River, South-central Kansas, 1995–99. USGS Water Resources Investigations Report 00-4126, 36 pp. http://ks.water.usgs.gov/pubs/reports/wrir.00-4126.html
23. Christensen VG, Rasmussen PP, Ziegler AC (2002) Real-time water quality monitoring and regression analysis to estimate nutrient and bacteria concentrations in Kansas streams. Water Sci Technol 45(9):205–211
24. Ryberg KR (2006) Continuous water-quality monitoring and regression analysis to estimate constituent concentrations and loads in the Red River of the North, Fargo, North Dakota. USGS Scientific Investigations Report 2006-5241, 35 pp.
25. Stubblefield AP, Reuter JE, Dahlgren RA, Goldman CR (2007) Use of turbidometry to characterize suspended sediment and phosphorus fluxes in the Lake Tahoe basin, California, USA. Hydrol Process 21(3):281–291
26. Berthouex PM, Brown LC (2002) Statistics for environmental engineers, 2nd ed. Lewis Publishers, New York, 489 pp.
27. http://mdw.srbc.net/remotewaterquality/methods.htm
28. WSN National Instruments. http://www.ni.com/wsn/
29. Manning EQUATION. http://www.iso.org/iso/home/store/catalogue_c/catalogue_etail.htm?csnumber=5564&commid=51675
30. USGS: Water Resources of Illinois: n-values Project (2012) http://il.water.usgs.gov/proj/nvalues/db/sites/05551540.shtml
31. Coulouris GF, Dollimore J, Kindberg T (2005) Distributed systems: concepts and design. Pearson Education
32. Buchanan BG, Duda RO (1983) Principles of rule-based expert systems. Adv Comput 22:163–216
33. Kone MT, Shimazu A, Nakajima T (2000) The state of the art in agent communication languages. Knowl Inf Syst 2(3):259–284
34. Ani EC, Cristea VM, Agachi PS, Kraslawski A (2010) Dynamic simulation of somes river pollution using MATLAB and COMSOL models. Revista de chimie, Bucuresti 11:1108–1112
35. Sinnott RW (1984) Virtues of the Haversine. Sky Telesc 68(2):159

Chapter 9
A Survey on Privacy Enhancements for Massively Scalable Storage Systems in Public Cloud Environments

Gabriel-Cosmin Apostol, Luminita Borcea, Ciprian Dobre, Constandinos X. Mavromoustakis, and George Mastorakis

Abstract Increasing network speeds and storage resource requirements have led the vast majority of industry players to intensively use cloud as a commodity service. Although the advantages obtained by following this trend are obvious, a multitude of privacy concerns arise regarding the confidentiality and integrity of data. In order to overcome privacy issues, some cloud providers and third-party software applications are now offering solutions which are capable of performing on-the-fly data encryption, in a simplistic manner for their clients. When it comes to encryption, there are a set of elements which should be known by the rightful owner, including the key, which should be kept secret, and the algorithm, which should be public in order to be reviewed. There are use cases when the secret key is managed by the client of the service and other use cases when the key is managed by the cloud provider. However, both generic use cases are subject to compromise due to viruses or human errors. Another security risk which arises for companies is caused by the fact that a single entity with access to the key could compromise trade secrets, thus causing irremediable loses.

Keywords Massively scalable storage system · Privacy enhancements for cloud environments · Mobile privacy overlays

G.-C. Apostol (✉) · L. Borcea · C. Dobre
University Politehnica of Bucharest, Str. Splaiul Independentei, nr. 313, Bucharest, Romania
e-mail: apostol.gabriel.cosmin@gmail.com

L. Borcea
e-mail: borcea.luminita@gmail.com

C. Dobre
e-mail: ciprian.dobre@cs.pub.ro

C. X. Mavromoustakis
University of Nicosia, 46 Makedonitissas Avenue, 1700 Nicosia, Cyprus
e-mail: mavromoustakis.c@unic.ac.cy

G. Mastorakis
Technological Educational Institute of Crete, Estavromenos, 71500 Heraklion, Crete, Greece
e-mail: gmastorakis@staff.teicrete

© Springer Nature Switzerland AG 2021
F. Pop and G. Neagu (eds.), *Big Data Platforms and Applications*,
Computer Communications and Networks,
https://doi.org/10.1007/978-3-030-38836-2_9

9.1 Introduction

Cloud computing is an emergent technology which tends to replace traditional computing and provides its users the means of storing their files on geographically dispersed data centers, with reduced costs or even for free. Cloud storage has become the preferred solution for small and even medium-sized companies in matters of cost-saving, collaboration, project management and productivity enhancement. The benefits of this technology are dimmed, however, by various security concerns regarding the cross-domain nature of the storage services. In order to comply with security policies and the legislation, various encryption techniques must be employed, in order to protect valuable company or personal assets.

Multiple solutions have been developed by storage providers or third parties, each having its own advantages in various use cases, thus having an important role in the adoption of secure storage cloud services. However, there are only a few providers which allow their clients to use their own keys and encrypt their data before uploading it to the cloud storage. Therefore, in the last years, a set of cloud storage aware privacy models were proposed in [3, 11, 18, 24, 34], providing additional security.

The proposed encryption schemes include key policy attribute-based encryption, ciphertext policy attribute-based encryption, ciphertext policy attribute set-based encryption, fuzzy identity-based encryption, hierarchical attribute-based encryption, hierarchical attribute set-based encryption and hierarchical identity-based encryption.

However third-party applications like Boxcryptor, Viivo and Cloudfogger overcome cloud privacy issues, by applying encryption overlays over insecure storage platforms.

9.2 Cloud Storage Encryption Prerequisites

In order to reduce traditional storage costs, some companies have started to outsource data storage to third parties to a certain extent. Even if all data should be the subject to storage outsourcing, we shall focus on sensitive data and how should it should be managed, in order to safe keep industry assets.

According to intellectual property protection methodology, there are four approaches a company could follow in order to protect its business assets, in order to maintain an economical advantage on the market.

The first three approaches imply the attachment of a **copyright** [21], a **trademark** [29] or a **patent** [32], which represent means of legal protection against replication.

The last approach implies the protection of assets using the **trade secret** methodology in order to prevent competitors to gain knowledge gathered from company-funded research and development.

Table 9.1 Advantages and disadvantages of intellectual property protection methods

	Trade secrets	Other means
Advantages	Unlimited protection time	Simplified licensing procedures
	Publication is not necessary	Grants exclusive rights opposable to anyone, including independent inventors
	Suitable for unprotectable work	
Disadvantages	Loss has immediate effects	Limited protection time
	No protection against independent inventors	Once published can be subject to research and development by competitors
	Some products are subject to reverse engineering	Reproducible by competitors after expiration

While each intellectual property mean of protection has its own advantages and disadvantages according to Table 9.1, we can observe the trade secrets could be protected using strong privacy cloud storage overlays.

According to Cundiff [5], an entity should constantly take reasonable measures when it comes to trade secrets protection in the digital environment. Even if not all trade secrets are subject to digitization, there are situations when a collaborative environment or an automated system imposes it.

A study published online in 2007 [22] stated that 60% of the questioned companies have estimated the amount of financial losses over 500.000 US dollars in the event of a security breach. Therefore, in the following years companies have started acknowledging the importance of security and begun the adoption of data encryption strategies as a means of protection suitable both for data communications and storage services.

However, technology can also pose a threat for company assets when not handled appropriately or handled by malevolent personnel. A company should define clear boundaries in the security policy regarding access areas, recording devices, communication systems and information for visitors, business partners and employees.

In order to overcome risks, an organization should constantly run training sessions in order to increase personnel awareness regarding trade secrets protection, especially when it comes to technology workers, which handle sensitive information inside the company.

There are situations when trade secrets are partially or fully disseminated to trusted business partners. In this case, the revealed information must be the subject of a non-disclosure agreement between the company and the third party, focusing on the importance and value of the shared knowledge.

9.3   Scalable Cloud Storage Encryption Schemes

The advances in cloud computing and the growing number of threats coming from the digital environment have led to the creation of many schemes used to ensure data security, confidentiality and access control. For a user to be able to share valuable information with a peer, the last one should be able to comply with a certain set of rules regarding access control.

The purpose of this chapter is to identify the existing cryptography-enforced access control schemes using Attribute-Based Encryption (ABE) in order to prevent intellectual property theft.

Traditional Public Key Encryption (PKE) systems would require a large number of keys, a complex management infrastructure and multiple copies of a file, each encrypted with a different key for each recipient [12]. In large companies, the usage of the PKE systems is impaired by a larger user base, requiring a newer, more scalable approach.

According to Goyal [9], one possible solution is to encrypt data using a set of attributes, enabling thus a one-to-many encryption scheme, suitable for preventing data duplication, thus reducing key management system complexities and storage costs. Another advantage resulted from the adoption of ABE consists of a strong protection means against collusion.

Attribute-Based Encryption can be structured in two main groups, namely Ciphertext Policy Attribute-Based Encryption (CP-ABE) and Key Policy Attribute-Based Encryption (KP-ABE). The main difference between the two schemes is the way that access policy is stored in order to prevent unauthorized access.

The CP-ABE scheme usage implies the association of a set of attributes with a private key belonging to a user and the embedding of an access policy regarding the system in the ciphertext. Decryption would be possible for an entity only when the owned attributes satisfy the policy terms. Policies can be created from a collection of attributes chained together using the logical operator and, the logical operator or and a (k, n) threshold rule, meaning that k out of n attributes must be satisfied in order to allow access. As an example, a company has three departments, which can be expressed as a universe of attributes $\{IT, HR, SL\}$. An executive, B, receives a key matching the attributes $\{IT, HR\}$ and an employee, A, receives another key matching the attribute $\{SL\}$. In order to grant access to both A and B to the same ciphertext, a policy formulated like $((IT \wedge HR) \vee SL)$ has to be employed before encryption. In order to allow the access for user A and forbid access for user B to a ciphertext, the access policy rule would have to be changed to $(IT \wedge HR)$.

Ciphertext Policy Attribute-Based Encryption is allowing users to share their data in a secure manner by specifying authorization rules which can be verified implicitly after the encryption process. The intended recipients can decipher data implicitly if their attributes match. Another scalable feature of this approach is given by the fact that users can encrypt data according to a policy, without having to worry who the persons are who will access it.

Key Policy Attribute-Based Encryption resembles the CP-ABE scheme in purpose, with the difference that its usage implies the association between a set of attributes and a private key belonging to a user. The access policy is embedded in the generated key, unlike in the CP-ABE scheme, where it resided in the ciphertext. The resulting encrypted information could only be accessed by a user with attributes accepted by the access control rule.

Ciphertext Policy Attribute-Based Encryption is a suitable scheme in a scenario where an individual needs to specify the access policy controlling a piece of information. CP-ABE is suitable for medical applications and social networks.

Key Policy Attribute-Based Encryption is a suitable scheme in a scenario where organizational policies impose a specific informational flow, which could be achieved by controlling the user-supplied keys. KP-ABE is suitable for event logging systems and database connection controls.

Each described ABE scheme requires a user proving its identity to a trusted authority before any decryption operation takes place. Even if the enumerated algorithms do not support multiple authorities, there are research directions focusing on multiauthority ABE schemes [1, 2, 19]. Another desirable set of features of ABE are accountability and attributes revocation, which are subject to further research [23].

9.4   Technology Survey Regarding Service Providers

File synchronization and backup over cloud platforms have become more and more ubiquitous in the last years. There are large companies which are offering services like Dropbox, SkyDrive, Google Drive, Box and Amazon Cloud Drive who are now offering both free and paid storage to users and companies. In order to observe the benefits of each platform, we performed a comparative analysis of the above enumerated services, based on white papers and online sources [4, 10, 13].

Cloud storage service solutions like Dropbox, Google Drive and Amazon Cloud Drive are convenient means of data backup and synchronization (Table 9.2). However, due to the fact that the data at rest is encrypted by the same storage companies, the services could access the files at any given time. A well-known fact is that each

Table 9.2 Comparison of storage cloud services

	Dropbox	Google Drive	Amazon Cloud Drive
File size limit	10 Gb	5 Tb	2 Gb
Encryption of data in transit	SSL TLS [2048 bit keys]	SSL TLS [2048 bit keys]	SSL TLS [2048 bit keys]
Free storage	2 Gb	15 Gb	None
Encryption of data at rest	AES [256 bits keys]	AES [128 bits keys]	None

cloud storage service provider has a policy regarding objectionable content, and might delete files at any given time without prior acknowledgment, even if the user has opted for a paid plan.

Another problem affecting the public cloud storage is the compromise of the user accounts, which would expose sensitive files to the attackers. Data would be accessible because encryption procedures are handled by the server and decryption is transparent for authenticated users.

9.5 Technology Survey Regarding Classic and Emerging Cryptographic Primitives

In order to obtain a better cost/security ratio, it is necessary to analyze the time and computational resource consumption for the classic and other emerging cryptographic primitives.

One popular metric unit for measuring Central Processing Unit (CPU) consumption is Cycles Per Byte (CPB), indicating the total number of clock cycles used by a microprocessor in order to perform one instruction of data—Cycles Per Second (CPS), calculated per byte of data processed—Bytes Per Second (BPS).

$$CPB = \frac{CPS}{BPS} \tag{9.1}$$

9.5.1 Confidentiality Primitives

Modern cryptography defines two types of symmetric encryption algorithms.

A cipher implies the existence of two paired functions (E—encryption and D—decryption), which accepts two inputs (n bits plain text and K—k bits encryption or decryption key) and provides an n bit encrypted or decrypted output text (Fig. 9.1).

$$E_K(P):=E(K, P) : \{0, 1\}^K \times \{0, 1\}^n \to \{0, 1\}^n (Encryption) \tag{9.2}$$

$$E_K^{-1}(C):=D_K(P) = D(K, C)\{0, 1\}^k \times \{0, 1\}^n \to \{0, 1\}^n (Decryption) \tag{9.3}$$

A stream cipher implies a Pseudo Random Number Generator (PRNG) to generate the key stream which will be used, with the input plain data stream, in the XOR function (Fig. 9.2). One such algorithm is the RC4 [26] stream cipher, which is defined by the following formulas, where P is a plain text bit, C is the cipher text bit and b is the generated key bit.

$$C = P \oplus b \tag{9.4}$$

$$P = C \oplus b \tag{9.5}$$

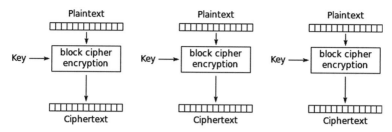

Fig. 9.1 Block ciphers encryption in electronic code book mode

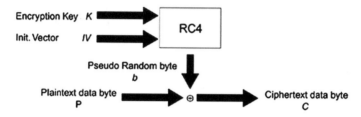

Fig. 9.2 RC4—stream cipher encryption implementation

Table 9.3 Encryption algorithm CPU consumption

Algorithm	Number of rounds	Block size (bits)	Cycles per byte	CPU Cycles to setup key and iv
DES	16	64	54.7	1532
IDEA	8	64	49.9	1277
AES	10/12/14	128	12.6	1277
RC5	12/16	32/128	23.4	4665
RC6	20	128	17.3	5128
MARS	2 × 16	128	37.2	6435
XTEA	32	32	67.4	1165
Serpent	32	128	54.7	2191
Twofish	16	128	29.4	14,121

In order to determine which of the modern cryptographic algorithms are suitable in our use case, there is necessity to evaluate the trade-off between CPU consumption, speed and security [6, 7, 14, 17, 20, 27, 28, 30, 31, 33]. Table 9.3 displays more details about each cryptographic primitive, although, in our critical analysis, we used only the data from the Number of Rounds, CPU Cycles per Byte and CPU Cycles to Setup Key and IV.

9.5.2 Integrity Primitives

Hash functions [8, 14, 25, 35] are usually used because can produce a short and fixed size digest output from a long, variable size input message. Also, these algorithms can assure a certain level of trust by providing a standardized mean to perform integrity checks on the received data.

From a cost perspective, the computational resource consumption is a metric equally important to the level of trust provided by the selected hashing method in power-constraint environments (Table 9.4).

From the analyzed data regarding hashing primitive performance, it can be determined that the smallest CPU consumption is obtained using the MD5 function. This approach could not be implemented in a digital signature scheme due to security concerns. A proper choice from a security standpoint would be a SHA-2 type of hash function. Another relevant factor is the time consumption required by the DSA and RSA algorithms (Table 9.5), due to the fact that each message should be signed in order to perform integrity and non-repudiation checks [14, 16].

Table 9.4 Hashing algorithms CPU consumption

Algorithm	Number of rounds	Input size (bits)	Hash size (bits)	CPU cycles per byte
MD5	4	512	128	6.8
SHA-1	4	512	160	11.9
SHA-256	64	512	256	15.8
SHA-512	80	512	512	17.7
WH	10	64	64	30.5

Table 9.5 Digital signature algorithm time consumption

Algorithm	Key size (bits)	Milliseconds per operation	Megacycles per operation
DSA 1024 bit signature	320	0.45	0.83
DSA 1024 bit verification	320	0.52	0.94
RSA 1024 bit signature	1024	1.48	2.71
RSA 1024 bit verification	1024	0.07	0.13

9.6 Technology Survey Regarding Third-Party Applications

While many individuals have already embraced cloud storage, new security concerns are beginning to rise in the aftermath of several high-profile data breaches [15] targeted toward the major players in the cloud storage industry.

In order to minimize the impact of a system breach, security experts all over the world have agreed upon the fact that encryption is the most suitable defense against cybernetic threats.

Unfortunately, encryption is not a silver bullet, and can be defeated if the systems depending on it are not implemented and managed properly. Every cipher is vulnerable to key loss or theft, if the key is poorly managed.

When it comes to cloud storage, the keys are managed either by providers, subscribers or by all the parties involved in the process.

Most cloud storage providers are implementing their own key management systems, reducing access complexity for the user. However, if an account gets compromised, there is no barrier in stopping an adversary to get the information stored using that account.

To overcome the issue of account compromise, independent researchers and security companies have started to create overlay systems for cloud storage platforms, ensuring data federation, confidentiality and integrity.

The most advertised apps in the domain of cloud storage security overlays are Boxcryptor, Cloudfogger, AES Crypt, SpiderOak and Viivo. These applications help their users to protect their data confidentiality before it reaches any server, adding an additional layer of security (Fig. 9.3).

In order to access the victim files, an attacker would have to know which overlay system was used, the key used by the overlay and the storage account of the account used for cloud synchronization.

9.6.1 Viivo

Viivo is a FIPS-140-2 validated software implementation which is free for personal use. It was created by the PKWARE company and integrates with Dropbox, Box, OneDrive, Google Drive and Copy.

It implements a Zero-Knowledge privacy policy, meaning that not even the cloud storage provider knows anything regarding the informational content of the shared files.

However, Viivo does not split encrypted files, meaning that sometimes, an attacker can still make a guess regarding the encrypted data file type based on its size. Another feature which is not enabled by default is file name encryption. Without file name encryption, an adversary could learn more on the contents of the file using the inference.

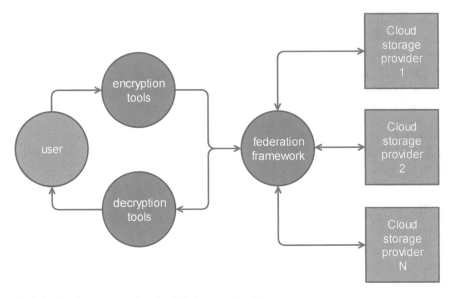

Fig. 9.3 Cloud storage overlay simplified encryption flow

The cryptographic primitives used by Viivo are RSA-4096 and AES-256, used together in order to achieve a higher degree of confidentiality (Fig. 9.4).

The encryption process (Fig. 9.4) is fully transparent for the users. The implementation of the encryption relies on a hybrid encryption approach, meaning that it uses both symmetric and asymmetric cryptographic primitives. Each file is encrypted with a different key, using the AES-256 algorithm. Each different key used for symmetric encryption is further encrypted asymmetrically with an RSA public key and the resulting cipher text is attached to the encrypted file. The decryption process works in reverse order, with the difference that, in the asymmetric decryption, the private key is used instead of the public one.

Viivo also supports file sharing, but the recipient must also use Viivo in order to be able to decrypt the file. This might be inconvenient, but it's a known fact that security is always associated with a lower degree of convenience.

Another feature not implemented in Viivo is a dual password mechanism in order to ensure plausible deniability. When an authority forces a Viivo subscriber to disclose his encrypted storage, such a mechanism would prevent him or her to disclose the real files, presenting instead a set of ordinary items.

Viivo can be used on multiple operating systems, like Apple Mac OS X, Windows and Android, ensuring cross-device interoperability and integration.

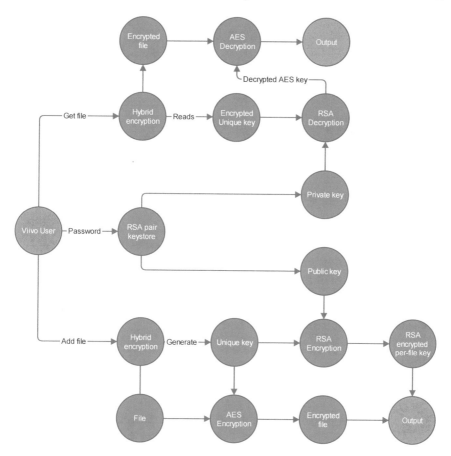

Fig. 9.4 Viivo storage overlay encryption data flow

9.6.2 AES Crypt

AES Crypt is a software implementation available for iOS, Android, Linux, PHP and Java that uses the standard Advanced Encryption Standard and Keyed-hash message authentication code using the cryptographic primitive Secure Hash Algorithm 2 with a 256-bit key to ensure data confidentiality and integrity.

Even if it does not implement any type of federation, this application is very simple to use across many devices and it has an open-source implementation, suitable for peer review.

The encryption process (Fig. 9.5) can be started by the user after a file and a password have been chosen. Regardless of the password complexity, the Password-Based Key Derivation Function 2 algorithm is used in order to derive the AES algorithm key and iv, and the HMAC-SHA-256 key.

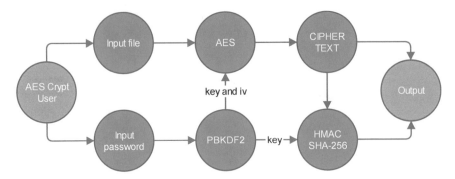

Fig. 9.5 AES crypt storage overlay simplified encryption flow

After the encryption process has been completed, the user has to manually upload his file to the cloud storage of his choice. In order to perform a file-sharing operation, a shared secret has to be previously established in a procedural manner.

Even if the contents of the files are encrypted, their file names are not, and could reveal informational content. However, if a user is careful enough, he can choose a safe naming scheme, in order to overcome this issue.

9.7 Proposed Solution

In this chapter, we propose a novel smart phone-based cloud storage encryption overlay, resilient to key theft, trojans, keyloggers, inference and account compromise. First we will describe the architecture of the system and then other relevant aspects regarding the functionality.

9.7.1 Architecture

One of the most sensitive aspects when it comes to cryptographic systems is key management, because there is no secure place to store keys in an Internet connected computer. One major issue in this case is malware, a family of hostile and intrusive software implementations targeting sensitive data. Another issue affecting confidentiality is the process of swapping, which cannot be stopped and could dump pieces of sensitive data to disk.

In order to overcome these issues, we propose the migration of sensitive data to another trust domain. This trust domain is a programmable smart card, enhanced with a cryptographical co-processor and powered by electromagnetic induction.

While a smart card requires additional setup and hardware, it is a small cost in order to achieve a higher degree of security for sensitive data.

9.7.2   General Description

The proposed cloud storage overlay will be capable of handling a file system-like structure in a manner that will not disclose the actual contents, file sizes or file names to an adversary, by performing encryption and decryption of fixed-size data structures stored on public infrastructures.

The solution will consist of two applications, a Java applet which would run on a Java Card and an Android application which will handle communication with the cloud storage and the installed applet.

9.7.3   The Java Card Applet

Java Card is a technology targeting low-power embedded devices in order to allow them to securely run applications. The card ecosystem is very secure, forbidding applet disclosure and tampering, in order to comply with various policies imposed by banks and governments.

Our proposal (Fig. 9.6) regarding the applet implies a software implementation which is protected by a pin and securely stores one key pair in order to perform data encryption for the files which will be uploaded to a public or private cloud storage.

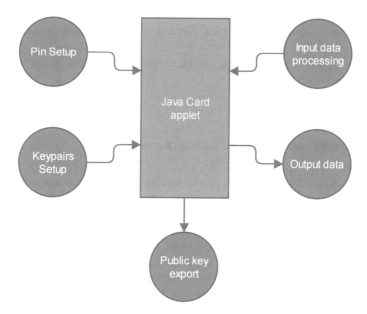

Fig. 9.6 Java Card applet architecture overview

The communication with the card will be made through NFC in order to perform applet deployment, PIN authentication, symmetric encryption, symmetric decryption, integrity computations, integrity checks, asymmetric encryption and asymmetric decryption.

The **pin setup** phase will be initialized in the moment when the card applet will be deployed, by setting the pin as an applet install parameter. The PIN can be changed later, but only after a successful authentication with the old PIN. If a Personal Identification Number code is typed incorrectly more than three times, the applet will lock itself and wipe the private key, preventing a brute force attack.

The **keypairs setup** phase will be initialized after the PIN code has been set up. The keys will not be generated on card, due to various limitations imposed by vendors. One of this limitations is caused by the lack of support for cryptographically secure random number generators. Another argument for this choice would be the fact that, if the card is lost, a new card would be reissued with the same key pair restored from a backup system.

The **public key export** is a feature which will enable users to receive encrypted files from other persons.

The **input data processing** will generate **output data** suitable either for cloud storage, if an encryption operation has been issued, either for visualization, if a decryption operation has been issued. The encryption flow will consist of

- A random 256-byte session key (K_S) will be generated.
- A hash function will be used in order to obtain a 48-byte hashed session key (HK_S).
- Using asymmetric encryption, (HK_S) will be encrypted and sent to the external domain (EHK_S).
- Using symmetric encryption, the input file will be encrypted with the (HK_S) key and sent to the external domain.

9.7.4   Storage Layout and Data Structures

The storage layout (Fig. 9.7) of our project will be implemented in a manner that will not allow an adversary learn information about file names, sizes and content.

In order to achieve previously mentioned characteristics, a data federation layer has to be implemented, in order to achieve integration with multiple storage providers. The federation layer will be connected to the hardware cryptographic interface, in order to perform secure read and write operations.

The storage layout will be split in four distinct zones, the metadata store, the temporary metadata store, the block store and the temporary block store. This separation will ensure file splitting, file name encryption, consistency and data encryption.

The **metadata store** will contain encrypted **metadata blobs** holding data regarding real file name, file size and block identifiers. The **metadata blob** file names will

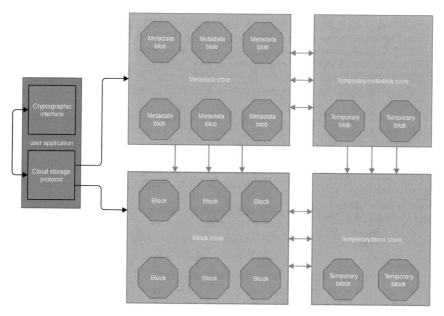

Fig. 9.7 Cloud storage architecture overview

be computed with a hash function applied over a timestamp obtained from an NTP server and a random number generated by application.

The **temporary metadata store** is an intermediate location used to securely store intermediate **metadata blobs** during the encryption process.

The purpose of the **block store** is to hold individual files containing secure information resulting from file encryption and splitting operation. The blocks will consist of files with random file names, in order to prevent real file names disclosure.

In order to ensure consistency, until the file encryption operation has been completed, the intermediate data will be uploaded to the **temporary block store**.

After an encryption operation has completed successfully, the data is moved from the temporary stores to the default stores. Any temporary file will have its name generated randomly.

References

1. Chase M (2007) Multi-authority attribute based encryption. In: Theory of cryptography. Springer, pp 515–534
2. Chase M, Chow SS (2009) Improving privacy and security in multi-authority attribute-based encryption. In: Proceedings of the 16th ACM conference on computer and communications security. ACM, pp 121–130

3. Chow R, Golle P, Jakobsson M, Shi E, Staddon J, Masuoka R, Molina J (2009) Controlling data in the cloud: outsourcing computation without outsourcing control. In: Proceedings of the 2009 ACM workshop on cloud computing security. ACM, pp 85–90
4. Corporation V (2015) Dropbox encryption vs. google drive encryption: which is more secure? https://www.virtru.com/blog/dropbox-encryption/. Accessed 12 June 2016
5. Cundiff VA (2008) Reasonable measures to protect trade secrets in a digital environment. Idea 49:359
6. Daemen J, Rijmen V (2013) The design of Rijndael: AES-the advanced encryption standard. Springer Science & Business Media
7. Dai W (2009) Crypto++ 5.6.0 benchmarks. Website at https://www.cryptopp.com/benchmarks.html
8. FIPS (2001) 180-2: secure hash standard (SHS). Technical report, National Institute of Standards and Technology (NIST), 2001. http://csrc.nist.gov/publications/fips/fips180-2/fips180-2withchangenotice.pdf
9. Goyal V, Pandey O, Sahai A, Waters B (2006) Attribute-based encryption for fine-grained access control of encrypted data. In: Proceedings of the 13th ACM conference on computer and communications security. ACM, pp 89–98
10. hinfinitsh (2015) Amazon cloud drive. https://infinit.sh/documentation/comparison/amazon-cloud-drive. Accessed 12 June 2016
11. Hwang JJ, Chuang HK, Hsu YC, Wu CH (2011) A business model for cloud computing based on a separate encryption and decryption service. In: 2011 international conference on information science and applications (ICISA). IEEE, pp 1–7
12. Ibraimi L, Petkovic M, Nikova S, Hartel P, Jonker W (2009) Ciphertext-policy attribute-based threshold decryption with flexible delegation and revocation of user attributes (extended version)
13. Inc D (2015) Dropbox business security (white paper). https://www.dropbox.com/static/business/resources/Security_Whitepaper.pdf. Accessed 12 June 2016
14. Kaps JP (2006) Cryptography for ultra-low power devices. PhD thesis, Worcester Polytechnic Institute
15. Khalid U, Ghafoor A, Irum M, Shibli MA (2013) Cloud based secure and privacy enhanced authentication and authorization protocol. Procedia Comput Sci 22:680–688
16. Kravitz DW (1993) Digital signature algorithm. US Patent 5,231,668
17. Lai X, Massey JL (1990) A proposal for a new block encryption standard. In: Workshop on the theory and application of of cryptographic techniques. Springer, pp 389–404
18. Li M, Yu S, Zheng Y, Ren K, Lou W (2013) Scalable and secure sharing of personal health records in cloud computing using attribute-based encryption. IEEE Trans Parallel Distrib Syst 24(1):131–143
19. Lin H, Cao Z, Liang X, Shao J (2010) Secure threshold multi authority attribute based encryption without a central authority. Inf Sci 180(13):2618–2632
20. Needham RM, Wheeler DJ (1997) Tea extensions. Report, Cambridge University, Cambridge, UK (October 1997)
21. Nimmer D (2013) Nimmer on copyright. LexisNexis
22. Norall S (2007) The growing importance of storage security and key management in large enterprises. https://www.globalsecuritymag.fr/article-Special-Reports,20071001,27.html. Accessed 12 June 2016
23. Pang L, Yang J, Jiang Z (2014) A survey of research progress and development tendency of attribute-based encryption. Sci World J
24. Park N (2011) Secure data access control scheme using type-based re-encryption in cloud environment. In: Semantic methods for knowledge management and communication. Springer, pp 319–327
25. Rivest R (1992) The MD5 message-digest algorithm
26. Rivest R (1992) The RC4 encryption algorithm, RSA data security inc. This document has not been made public

27. Rivest RL (1994) The RC5 encryption algorithm. In: International workshop on fast software encryption. Springer, pp 86–96
28. Rivest RL, Robshaw M, Sidney R, Yin YL (1998) The RC6TM block cipher. In: First advanced encryption standard (AES) conference
29. Schechter FI (1927) The rational basis of trademark protection. Harv Law Rev 40(6):813–833
30. Schneier B, Kelsey J, Whiting D, Wagner D, Hall C, Ferguson N (1998) Twofish: a 128-bit block cipher. NIST AES Proposal 15
31. Standard DE (1977) FIPS publication 46
32. US Patent and Trademark Office UDoC (2001) Manual of patent examining procedure
33. Verma OP, Agarwal R, Dafouti D, Tyagi S (2011) Performance analysis of data encryption algorithms. In: 2011 3rd international conference on electronics computer technology (ICECT), vol 5. IEEE, pp 399–403
34. Wang G, Liu Q, Wu J, Guo M (2011) Hierarchical attribute-based encryption and scalable user revocation for sharing data in cloud servers. Comput Secur 30(5):320–331
35. Yuksel K, Kaps JP, Sunar B (2004) Universal hash functions for emerging ultra-low-power networks. In: Proceedings of the communications networks and distributed systems modeling and simulation conference

Chapter 10
Energy Efficiency of Arduino Sensors Platform Based on Mobile-Cloud: A Bicycle Lights Use-Case

Alin Zamfiroiu

Abstract IoT technology is a platform where devices become more intelligent and make decisions in a very short time, bringing optimal solutions to problems in everyday life. These solutions are based on the previous behavior of certain systems or people. In this material, we intend to make a prototype smart device that contributes to efficient use of energy for lights that are equipped bikes. Enlightenment is based on street traffic on which the bicycle circulates. Since the device determines in real time the degree of agglomeration in the streets, such data may also be sent to the cloud for further analysis. This helps analyze traffic on public streets.

Keywords Mobile · Cloud · Arduino · Sensors · Energy · Architecture

10.1 Introduction

According to Trappey [1], "Internet-of-Things", or in short, IoT is a term used for the first time by Kevin Ashton in 1999, a well-known British pioneer in information technology. He used the term to describe an integrated system between the digital world (the Internet) and the real world.

According to [2], Internet-of-Things is a platform where all devices become smart, and from day to day they improve their behavior.

In [2], an analysis is made of the domains and subdomains in which IoT is used, they are presented in Fig. 10.1.

From these domains and subdomains within this material we are dealing with the Smart Society and the Smart Cycling sub-domain, more precisely, we are looking at the energy efficiency of the lights used by a bicycle.

A. Zamfiroiu (✉)
Bucharest University of Economic Studies, Bucharest, Romania
e-mail: alin.zamfiroiu@ici.ro

National Institute for Research and Development in Informatics, Bucharest, Romania

© Springer Nature Switzerland AG 2021
F. Pop and G. Neagu (eds.), *Big Data Platforms and Applications*,
Computer Communications and Networks,
https://doi.org/10.1007/978-3-030-38836-2_10

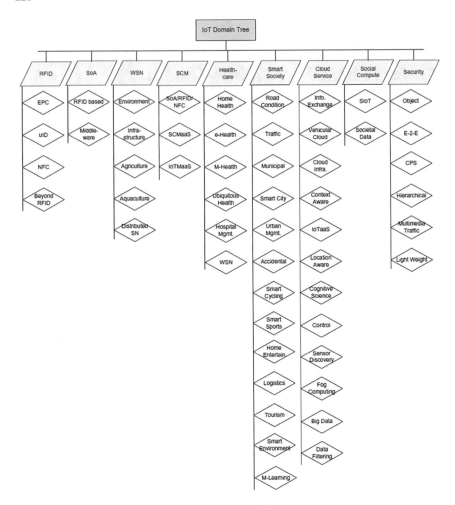

Fig. 10.1 Domains of application of IoT technologies [2]

This involves lighting them when needed, at the required intensity and frequency required by the traffic situation. In this way, the simple stop of the bike becomes a smart device [3].

There are many bicycle lighting systems that use as a saving technique, lighting several bulbs alternately. Battery life is prolonged.

In this paper, it is proposed to establish a system for determining the light need to be produced by the bicycle in traffic and lighting the lights to announce the participants about the existence of the bicycle-only when this is necessary. The proposed system brings optimal operation and reduces energy consumption according to existing traffic.

This stop while the bicyclist is alone and has no car in the back, will light up for a longer time because no car in the back is ignition unnecessary.

Fig. 10.2 The distance between the car and the bicycle

When a car approaches the cyclist, the stoplights more often depending on the distance between the car and the bike. The distance between the car and the bicycle is determined based on the light from the car headlights, Fig. 10.2.

In order to determine the power of the light from the car headlights, a light sensor is used, and real-time tests will establish intervals for calculating the distance to which the bicycle machine is located.

When the car is very close to the bicycle and the latter is very bright, the stop should not turn on and so it remains off until it has no car behind it. In this way, the efficiency is maximal.

10.2   Mobile Cloud Computing

In the beginning, the issue of IoT technology was the huge amount of data to collect and the cost of storing them. If we want to store information about the huge number of real-life objects, storage requirements were going to be huge. The solution came with Cloud technology that provided the data center infrastructure with much more storage and processing capacity than in the past [4–6].

The ESP8266 WiFi module, Fig. 10.3, is a microchip containing all circuits and electronic parts [7] with TCP/IP integrated protocol, capable of making WiFi communications and communicating with other Arduino boards or devices [8].

With the ESP8266 module, the data collected by the Arduino board will be transmitted in cloud. Because the intelligent stop detects the cars that pass by the cyclist, it will transmit in the cloud the number of cars that pass on the street in a time span [9]. This makes it possible to determine the degree of agglomeration of the street.

For cloud storage, *Blink tools* is used. Blink tools are:

- Blink app—used to create applications with interfaces for the developed projects;

Fig. 10.3 WiFi ESP8266 module

- Blynk Server—used for storage and communication between the hardware project and the mobile device; if the developer wants his or her own server, or wants to be stored on the mobile device, a mobile cloud is created; In this way, there is no need for a server or physical computer, but only for sensors, Arduino board, a WiFi module, and a mobile phone with the installed Blynk application, Fig. 10.4;
- Blynk Libraries—are the libraries used for communicating between sensors and a mobile server/phone.

Data sent to BlynkApp can then be used to produce statistics or make a history of the streets circulated by the cyclist, the hours he has circulated, and the number of cars that have passed by him at those times. In this way, the degree of agglomeration of the streets over a certain time interval is determined. No GPS sensor is needed to get the location, as the GPS module of the mobile phone can be used.

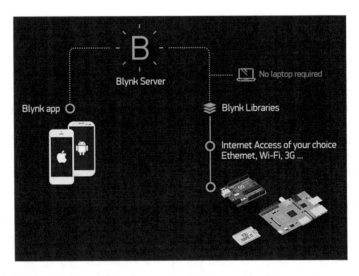

Fig. 10.4 Blynk architecture [10]

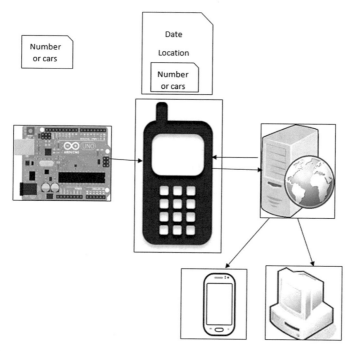

Fig. 10.5 The data transfer architecture in Mobile Cloud

Thus, the Arduino board only sends the calculated number of machines, then through the Blynk App application, new metadata is added such as location and date, represented by the day and time at which that information was received.

This data is sent to a server and can be accessed by another mobile device or by a computer with an Internet connection [11]. These applications also need to be developed.

10.3   The System for Energy Efficiency of Arduino Sensors

The following equipment is required to develop the prototype system:

- a bicycle stop light that can be changed;
- an Arduino development board;
- a brick light sensor;
- a led to turn on the stoplight;
- father-father interconnection wires.

The brick light sensor connects three wires to the Arduino board. An input thread, a thread for feeding, and a ground wire [12]. The LED connects to the Arduino board by two wires, one for earthing and one for receiving the ignition or extinguishing signal of the LED, Fig. 10.6.

Their interconnection with the Arduino board is done for the debug and the source code in accordance with the reality, and after it has been fully tested and tested to be copied to an Adafruit Trinket board and framing it in a bicycle stoplight.

To achieve the ignition thresholds to make energy consumption more efficient, measurements were made on a field where there is no other light source, but only the headlights of a car. Measures were taken at 500, 300, 100, 50, 30, 10, and 1 m. The data obtained with the brick light sensor, when the headlamps were illuminated over the long-range, are shown in Table 10.1.

Measurements were made at the same distances from the dipped-beam headlamps. These are presented in Table 10.2.

Considering the legal speed of 90 km/h, the vehicle's driving times over the central bicycle are calculated. Durations are shown in Table 10.3.

As the measured thresholds are close, only the following four thresholds, presented in Table 10.4, will be considered.

Fig. 10.6 Interconnection of electronic components

Table 10.1 Long-range measurements

Distance	Measured level
500	1
300	1
100	16
50	43
30	86
10	500
1	700

Table 10.2 Short-phase measurements

Distance	Measured level
500	0
300	1
100	15
50	40
30	80
10	500
1	700

Table 10.3 The average time to cross the distance

Distance	Duration (seconds)
500	20
300	12
100	4
50	2
30	1.2
10	0.4
1	0.04

Table 10.4 Considered thresholds

Distance	Threshold	Threshold value
300	Prag 1	1
100	Prag 2	15
50	Prag 3	40
10	Prag 4	500

Based on the values of these thresholds, the lighting ignition algorithm is developed so that the consumption is efficient.

10.4   Smart Bicycle Lighting Architecture

The energy efficiency program should accurately determine when the positioning light should be lit and at what frequency it should switch between the ignition and the off position.

For this, the program contains a part of declaring the variables used. In the declaration part, the experimental thresholds are used and the led port is set.

```
int interval = 150;
int prag1 = 1;
int prag2 = 15;
int prag3 = 40;
int prag4 = 500;
int multiplicator1 = 1;
int multiplicator2 = 10;
int multiplicator3 = 4;
int multiplicator4 = 1;
int ordin_schimbare_status = 10;
int index_aprindere = 0;
int status_led = 0;
int led_port = 8;
int timp_aprindere = 2;
```

These variables are used to determine the amount of time the LED should light up and the amount of time the led should be off.

In the setup section, the input ports for that light sensor, the debug port, are configured.

```
void setup() {
Serial.begin(9600);
pinMode(led_port, OUTPUT);
}
```

The loop section reads from the analog port of the value transmitted by the brick light sensor. This value is used to determine how the LED flashes.

The following mathematical function for determining the level of growth is used:

$$index = \begin{cases} ivl * mult1 \ val \leq prag1 \\ ivl * mult2 \ val > prag1 \&\&val \leq prag2 \\ ivl * mult3 \ val > prag2 \&\&val \leq prag3 \\ ivl * mult4 \ val > prag3 \&\&val \leq prag4 \\ 0 \ val > prag4 \end{cases}$$

$$index\_aprindere = index\_aprindere + index$$

where:

- iv—the interval at which the loop is repeated;
- mult1, mult2, mult3, mult4—experimentally determined multiplication values;
- val—the value read by the light sensor;
- prag1, prag2, prag3, prag4—the thresholds for the distance between the car and the bicycle, values obtained experimentally.

The loop section code implements the above formula. Thus, depending on the light behind the cyclist, the interval at which it stops illuminates.

Finally, if the ignition index value reaches the ignition threshold, the status of the LED changes, if the ignition is switched off, and if it is extinguished now is lit and the ignition index value reset.

The code is copied to Adafruit Trinket. The ports for communicating with the light sensor and the led to light are changed.

For the Trinket card, it is also needed mother-mother wires that are connected to those father–father from the sensor and led.

Both the led and the plate are inserted into a bicycle stop and the light sensor is kept out of the stop, making a hole for inserting its threads for communication with the Trinket board, Fig. 10.7.

The stop is made by a new hole through which to feed the Adafruit Trinket board. The power is supplied from an external battery, Fig. 10.8.

Fig. 10.7 Connect the sensor and led to the Trinket board

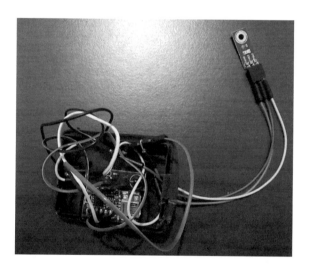

Fig. 10.8 Power supplier for the intelligent stoplight

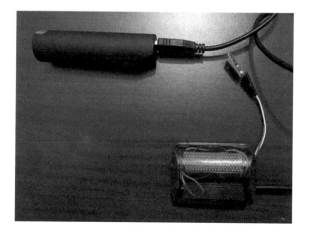

Fig. 10.9 Installation on the bicycle

The stoplight is mounted on the bicycle and the supply line is searched for enough that the battery is installed at an optimum distance from the stop and the light sensor is not disturbed by certain elements of the bicycle Fig. 10.9.

The battery can be changed at any time, and a larger battery can be used.

10.5 Conclusions

The prototype has been tested and works very well on the bicycle. In the future, we aim to achieve the smart stoplight with several LEDs and ignite them after the analyzes made for the data gathered by this prototype.

Based on the agglomeration level analysis on certain streets, we aim to create optimal routes for cyclists so that they will bypass the crowded streets. In this way, mobile technologies, IoT and Cloud platforms contribute to optimizing decisions on energy efficiency for bicycles.

References

1. Amy JC, Trappey CV, Govindarajan UH, Chuang AC, Sun JJ (2016) A review of essential standards and patent landscapes for the Internet of Things: a key enabler for Industry 4.0. Adv Eng Inform 22
2. Ray PP (2016) A survey on Internet of Things architectures. J King Saud Univ—Comput Inf Sci 29

3. Tang S, Kalavally V, Ng KY, Parkkinen J (2017) Development of a prototype smart home intelligent lighting control architecture using sensors on board a mobile computing system. Energy Build 138: 368–376
4. Liu Z-Z, Li S-N (2018) Sensors-cloud data acquisition based on fog computation and adaptive block compressed sensing. Int J Distrib Sens Netw 14(9)
5. Antonic A, Marjanovic M, Pripuzic K (2016) Ivana Podnar Zarko, A mobile crowd sensing ecosystem enabled by CUPUS: Cloud-based publish/subscribe middleware for the Internet of Things. Future Gener Comput Syst 56:607–622
6. Botta A, de Donato W, Persico V, Pescape A (2016) Integration of cloud computing and internet of things: a survey. Future Gener Comput Syst 56:684–700
7. Wikipedia ESP8266. https://en.wikipedia.org/wiki/ESP8266
8. Zafar S, Miraj G, Baloch R (2018) An IoT based real-time environmental monitoring system using Arduino and cloud service. Eng, Technol Appl Sci Res 8(4):3238–3242
9. da Silva AF, Ohta RL, dos Santos MN, Binotto APD (2016) A cloud-based architecture for the internet of things targeting industrial devices remote monitoring and control, IFAC, Conference Paper Arhive 49(30): 108–113
10. Blink documentation. http://docs.blynk.cc/
11. Santos J, Rodrigues JJPC, Silva BMC, Casal J, Saleem K, Denisov V (2016) An IoT-based mobile gateway for intelligent personal assistants on mobile health environments. J Netw Comput Appl 71: 194–204
12. Xia C, Zhang Y, Liu Y, Lin K, Chen J (2018) Path planning and energy control of wireless power transfer for sensor nodes in wireless sensor networks. Turkish J Electr Eng Comput Sci 26:2618–2632

Chapter 11
Cloud-Enabled Modeling of Sensor Networks in Educational Settings

Florin Daniel Anton and Anca Daniela Ionita

Abstract Interdisciplinarity is an important challenge for nowadays engineering studies and this can be achieved through systematic integration of concepts and technologies. The chapter presents an approach that gives an inner view on conceiving modeling languages with specific applications to sensor networks, supported by configurable tools enabled by cloud. The system is used by students to model the characteristics of the sensors and the network architecture, but also to introduce their extensions through programs that interpret such models. The modeling environment uses Windows as a host operating system and is deployed in a customizable Infrastructure as a Service cloud, based on IBM Service Delivery Manager. For achieving an easily deployable educational solution, a virtual machine is associated with each student who works to accomplish a task for a given laboratory class, or during the entire semester. The provisioning process and experimental results for several test scenarios are also described.

11.1 Introduction

Although initially applied in military scenarios, sensor networks are currently used at a large scale, for monitoring vehicles, buildings, and industrial plants, especially based on the adoption of the wireless solutions [12]. Their evolution was conducted by two trends that coexist, though reflect contradicting principles. On the one hand, the sensor network architectures have become increasingly distributed, with computations hosted by smart devices spread in multiple locations. On the other hand, they became even more centralized as the servers traditionally used for managing the collected information are often replaced by cloud computing platforms. A good balance between the two trends has been reached in the fog computing approaches,

F. D. Anton · A. D. Ionita (✉)
University Politehnica of Bucharest, Spl. Independentei 313, 060042 Bucharest, Romania
e-mail: anca.ionita@upb.ro

F. D. Anton
e-mail: florin.anton@upb.ro

© Springer Nature Switzerland AG 2021
F. Pop and G. Neagu (eds.), *Big Data Platforms and Applications*,
Computer Communications and Networks,
https://doi.org/10.1007/978-3-030-38836-2_11

combining distributed analytics with centralized monitoring and decision making [29].

In this context, our interest in cloud-enabled sensor networks stands in adding modeling capabilities to the sensing and processing ones and, more specifically, in supporting *Sensor Network Modeling as a Service(SN-MaaS)*, in addition to *Sensing as a Service* (S^{2}aaS) [31] and *Sensor Event as a Service* (SEaaS). We introduced a first example of MaaS application for sensor networks in [20], where we presented the deployment in cloud of a modeling environment for sensor networks, which makes a clear difference between sensing and processing units, and also gives support for an integrated view of them, with an example for the vehicle detection on a multiple lane roadway. The approach is different from the types of sensor networks based on Service-Oriented Architecture (SOA), as the one presented in [10].

This chapter will get into more insights on the processes applied, both for a detailed definition of the modeling elements for the sensor network editor, and for provisioning the modeling capabilities as a service, in a university private cloud relying on an IBM platform.

The remaining of the chapter is structured as follows: the next section presents related work regarding sensors in cloud and education in cloud, followed by a section concerning a modeling language, tools and model interpreters for sensor networks. Then, the chapter describes the architecture of the cloud system that hosts the modeling environment and continues with the educational service in cloud, seen from the point of view of service request and handling, and service provisioning. The chapter ends with experimental results, considering the query and cloning stages, as well as the hardware and software reconfigurations, and conclusions.

11.2 Related Work

11.2.1 Sensor Cloud

As sensor networks have become more accessible and capable to provide large-scale data as input for complex visualization and intelligent processing, cloud computing environments have been investigated as a solution to many situations. Some application domains where this mixture may be beneficial are: smart transportation, military surveillance, environmental monitoring for weather forecasting and disaster detection, health care, telematics, tunnel monitoring, wildlife observation [2, 9]. Chung et al. [7] discuss about Big Data originated from sensor nodes deployed in an agricultural system, sent to a distributed database in cloud; the storage capability would thus suffice for very large areas and even for an entire country. A framework to integrate sensors and cloud, including publish–subscribe brokers, was proposed in [1], with potential applications for wireless sensor networks of hundreds of nodes, for measuring drinking water, or environment conditions in a rainforest. An architecture composed of sensing nodes, decision nodes, and gateways in cloud was conceived for adapting the evacuation paths in case of emergency in respect with the building constraints, the hazard intensity, and the availability of external services for supporting the evacuation [3]. Cloud can also act as an intermediate environment between physical and virtual sensors, a new paradigm introduced in applications like mobile crowdsensing, or vehicular cloud computing [18].

11.2.2   Education Cloud

Education can also benefit from a large variety of cloud services [5], some of them being generic, like e-mails, search, portals, wikis, blogs, video conferencing, while others being much specific, like digital portfolios, electronic assessment, content access and creation, up to platforms and software that are dedicated to the study of a single discipline (Fogel 2010). The access to high-performance computing facilities and very expensive equipment (through remote experimentation by appointment) may also have a big impact on student learning [23]. Moreover, cloud is considered one of the enablers of Education 2.0, which introduces new educational forms, like personalized supervision, virtualization of desktops, collective teaching, involvement of parents, real-time assessment, teaching analysis, etc. (Applying the cloud in education 2012).

Cloud computing was identified as an opportunity in the conditions of internationalization of higher education, of increased students and staff mobility in Europe, raising the issue of interoperability, but also of a wider access to educational materials in developing countries [17]. However, several barriers of cloud adoption in public education are: lack of trust in remote storing of data, legislation incompatibilities for administration services, and discontinuities of the Internet access [25]. Permission for cross-boundary transmission of information, data privacy, and other ethical issues are also mentioned in [11]. Thus, a survey realized by ViON on colleges and research universities, ranging from 2500 to 10,000 students, reveals that most of the services use private clouds, and access to use was the most important benefit that was appreciated [32].

11.3   Sensor Network Modeling

11.3.1   Language and Tools

The sensor network modeling is performed with a language developed in-house, based on the core concepts of the domain and complying with terminologies available in the scientific literature. It separates the modeling views for communication, memory, power, sensors, and network elements, and supports a uniform way of representing sensor network architectures, to allow their comparison, evaluation, and eventually simulation [20]. The modeling tools for this language were realized with Generic Modeling Environment (GME) (2017) and they may be extended by adding supplementary views, or by developing services based on model interpretation.

The development cycle for creating these tools started with collecting multiple real-life sensor networks and taxonomies available for this domain. The examples were *analyzed* to extract a set of common concepts and relationships between them, general enough for describing any of the studied networks. Figure 11.1 represents a part of the terminology adopted, where the architecture types are originated from [14], the general aspects are those specified in [6] and the sensor characteristics and measurand types correspond to the classification from [33]. The terminology from Fig. 11.1, along with other domain-specific concepts, were used to *define a modeling language*. The resulted modeling elements were formalized in GME as illustrated in Fig. 11.2, in order to *create modeling tools*, which were deployed in the cloud environment to provide Sensor Network Modeling as a Service. The SN-MaaS tools were subsequently applied to *model the examples* of sensor networks that were previously analyzed. The overall development cycle for creating SN-MaaS tools is represented in Fig. 11.3.

The development cycle may be reiterated in the following situations:

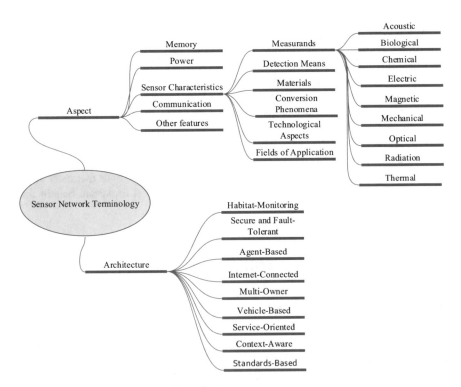

Fig. 11.1 Part of the sensor network terminology

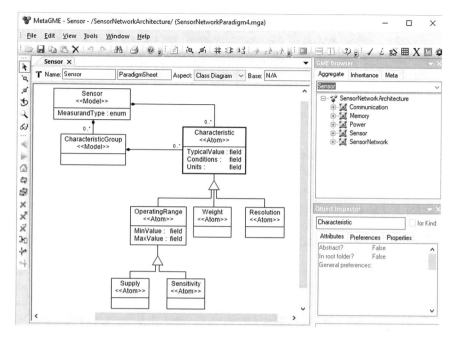

Fig. 11.2 Sensor network modeling elements formalized in GME

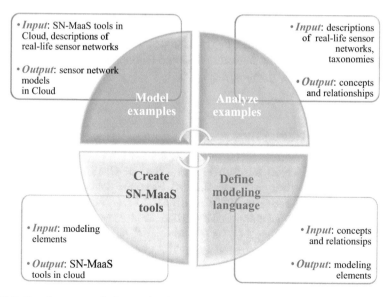

Fig. 11.3 Development cycle for creating SN-MaaS tools

- in case some elements of the already analyzed networks cannot be represented with the current version of the modeling tools, the cycle is re-initiated directly from the definition of the modeling language;
- in case a new set of real-life examples of sensor networks emphasize limitations of the current SN-MaaS tools, an in-depth analysis is necessary, and the cycle has to be re-initiated from the beginning.

11.3.2 Extensions and Model Interpreters

Starting from the current environment that offers *Sensor Network Modeling as a Service*, it is also possible to add other types of services, by developing model interpreters. This allows the evolution from the capability of modeling sensor network architectures to executing models and using their outputs in other programming contexts.

An example is an interpreter that transforms the graphical model into a sensor network specification in SensorML (OpenGIS 2007). The scopes of SensorML and our modeling language overlap, but they are not identical, so only a part of the modeling elements can be mapped to each other. Such an interpreter was developed based on an intermediate XML model and two transformations: the former realized by mapping the modeling elements of our language to a correspondent XML schema, and the latter by mapping the elements of the XML model to SensorML, with its specific markup elements [21].

The current modeling tools can also be extended with other services, for example, for detecting the sensors not delivering data and eliminating them from the model, or for automatically adding new sensors.

11.4 System Architecture

The system is based on an Infrastructure as a Service (IaaS) cloud, implemented using IBM Service Delivery Manager (ISDM). From the point of view of hardware, the ISDM solution is based on an IBM CloudBurst 2.1 medium deployment solution, which is a centralized cloud composed by a BladeCenter H rack with 14 HS22 Blade servers, each of them having 2 CPUs (6 cores per CPU) and 72 GB of RAM; the blade servers do not have internal storage. The system has also an IBM DS3400 storage with three expansions, totaling 48 SAS disks, each disk having 600 GB. The management of the IaaS is done using another server, which contains 2 CPUs (each CPU having 8 cores) at 2.4 GHz and 24 GB of RAM. The entire infrastructure is powered using 2 APC ups systems, each having 10KVA, which allows the system to shut down in a power failure situation [22, 24].

The ISDM solution was selected because it is more flexible than CloudBurst, allowing one to customize the hardware infrastructure; for example, one can add physical machines with different architectures (machines with Intel processors, Power machines, or mainframes) which run different hypervisors (vmware, xen, powervm, etc.) [19].

From the point of view of the software architecture, the system uses two management machines:

1. A standalone server, used to store the provisioning information and the status of each virtual machine; for this, an MS SQL Server database was used; the database is accessed directly by a VMware VSphere server that manages all the hypervisors (VMware ESXi 4.0) of the 14 blades installed in the BladeCenter H

2. A blade server that runs four virtual machines (IBM Tivoli Service Automation Manager—TSAM, IBM Tivoli Usage and Accounting Manager—TUAM, IBM Tivoli Monitoring—ITM, Network File System—NFS); each machine is installed with SuSE Linux and has the following functions:

- The TSAM machine is installed with SuSE Linux 10, TSAM, and Tivoli System Automation—TSA; this machine allows one to process service requests from clients for deploying virtual machines (VMs) [30] [26]. TSAM uses multiple components in order to offer services to clients. When a client is requesting a service, the client connects to a Self-Service web interface, where he or she can choose the desired service, and the time when the service should be provided, e.g., 10 VMs installed with a specified operating system (Windows or Linux) and a specified set of software applications, which should be ready at a specific date. This request is forwarded by the self-service interface, using REST API, to TSA, where a component called Tivoli Service Request Manager (TSRM) processes this request, by reserving the resources for the service request for the specified period of time. When the provisioning time comes, the TSRM component sends a number of provision requests, equal to the number of requested VMs, to another component, which is called Tivoli Provisioning Manager (TPM) and is responsible with the creation, configuration, installation, and decommissioning of the VMs [27].

- The ITM machine is responsible for acquiring monitoring data from the virtual machines, through two mechanisms: using agents deployed on each operating system on the VM, or using data provided from the VMware ESXi hypervisors. ITM gathers data like: CPU, RAM, storage, network bandwidth usage, power consumption, all of them being tagged with timestamps. The acquired information is placed into a DB2 database and offered to the TUAM VM [16].

- The TUAM machine is used to extract the data stored by ITM from the database, and to generate usage reports and payment information [8]. Because the system is an IaaS, the payment mechanism is "pay as you go"; this machine allows the clients to monitor their costs and also allows the service provider to generate invoices for each offered service.

- The NFS machine is installed with TSA, HTTP, NFS and Samba servers and is used to offer a software repository for the VMs. The software and the agents that are installed on VMs after the VMs are cloned and configured are stored on this machine. Moreover, this machine offers the HTTP service for the clients that access the Self Service interface.

Figure 11.4 presents the simplified hardware architecture, containing the Blade-Center with 14 HS22 Blade Servers, the network connectivity switches (1G SM Bay 1 and 2), the connectivity with the DS3400 storage system (2 Fiber channel switches—10pt FC SM Bay 3 and 4); the BladeCenter also has two advanced management modules (AMM 2). The server X3650 M2 is the management machine, running VMware vSphere server, and the first blade (on the left side) is running the four VMs (presented above) used for cloud service administration. The entire system is powered by two power distribution units (PDU).

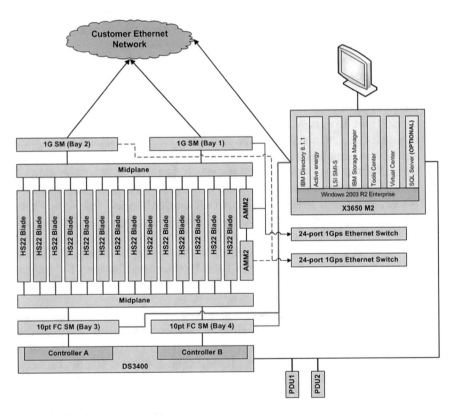

Fig. 11.4 The cloud system architecture

11.5 Educational Service in Cloud

11.5.1 Service Request and Handling

The service request and handling process are realized using the management machines. The service request is placed into the system by means of the Self-service GUI, using a simple web browser. Here, the client or user (student or teacher) must authenticate and then request a new service by creating a new project request (see Fig. 11.5).

A project contains a set of information, like project name, project start and end date, resource pool used to deploy the VMs, VM template that contains the operating system (the VM templates or the VM images available on the system are queried by using REST API), additional software that will be installed on the VMs (including the monitoring agent) and hardware configuration of the VMs (CPU, RAM, and HDD); the network is configured based on the selected pool of resources, each cloud pool having a customized network configuration settings, which is applied to the VMs deployed on it; the configuration specifies how many network adapters will be installed on the VM, what are the Virtual LANs used and the network card configuration parameters (IP, Netmask, Gateway and DNS).

After creating the project request (or service request SR), the SR is received by Tivoli Service Request Manager (TSRM), which allocates resources for the project (after the project is approved)—for the lifetime of the project—and keeps the request in a queue until the project start date.

When the start date arrives, TSRM creates a separate SR for each VM and sends a service definition to Tivoli Service Automation Manager (TivSAM). The service definition contains a service topology that contains the network settings of VMs, and the relationship between VMs and management plans for the VMs in the project; the

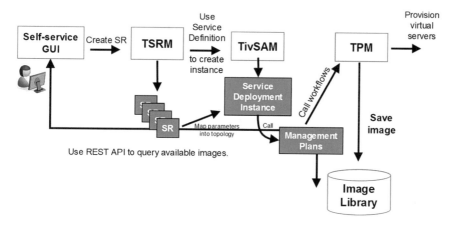

Fig. 11.5 Service request and handling

management plans contains the procedures that must be executed in order to manage the VMs (create the VM, add software, reset password, etc.).

TivSAM generates a service deployment instance that sends the management plans to Tivoli Provisioning Manager (TPM) in the form of workflows (step-by-step procedures) to execute the management plans. TPM executes a workflow in order to provision, modify, decommission or save the image of a VM in the image library.

11.5.2 The Provisioning Process

The provisioning process, from the point of view of user interaction, is explained in Fig. 11.6. The client or the user uses the self-service GUI to create a service request (a project that contains multiple VMs, having a specific hardware and software configuration). The request is sent to the cloud administrator, who approves the project.

After the approval, the service request is processed by TSAM, which allocates resources to the project and creates the VMs through TPM. After the creation of the VMs, the team members of the user who requested the service are notified by email and can access the VMs. If other users are later added to the team, the user who requested the project can make another request, to add supplementary VMs. The request is sent to TSAM, which creates the VMs, notifies the users, and the VMs can be accessed by the new team members.

From the point of view of the service lifecycle, the entire process is presented in Fig. 11.7. In the first step, the team administrator requests a project composed of a

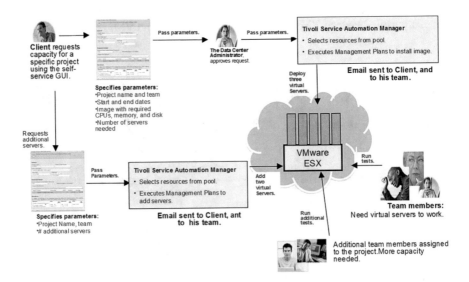

Fig. 11.6 The provisioning process

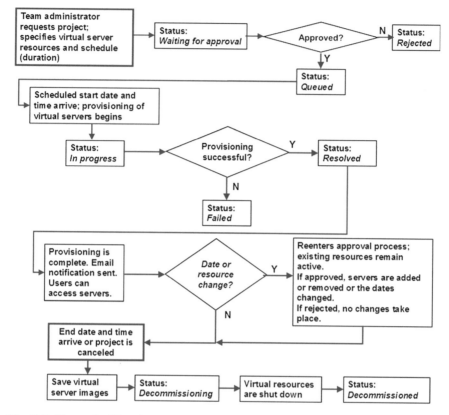

Fig. 11.7 The service lifecycle

number of virtual machines; after the request is submitted, the status of the project is "Waiting for approval". In this step, the cloud administrator is notified, and he or she must approve or reject the request. If the request is rejected, the status of the project becomes "Rejected", the service request is discarded, the user who placed the request is notified and no other processing is done. If the project is approved, the service request is placed into a queue and the system awaits the scheduled date for provisioning; during this time, the project status is "Queued". Irrespective of the fact that the project is approved or rejected, the user who requested the service will be notified about the decision of the cloud administrator.

When the scheduled start date and time arrive, the service request is processed and TPM starts to clone the VM template, in order to create and configure the VMs; in this step the status of the project is "In progress". If TPM fails to create the VMs, the status of the project becomes "Failed" and the cloned VMs are erased and the allocated resources for the project are freed and the user notified.

If the provisioning is successful then the project status becomes "Resolved", then the user who requested the service and his or her team members are notified on how they can access the VMs: IP addresses, connection method (i.e., Remote

Desktop Protocol—RDP for Windows, and Secure Shell—SSH for Linux), user and password; then, the project becomes "Operational".

If during the operational phase of the project the team administrator requests to add other servers to the project or to modify the project end date or the hardware resources for one or more VMs of the project, the request can be approved immediately by the system, based on some thresholds established on the system configuration; for example, if the RAM of a VM is requested to be raised under a threshold of 4 GB the request is automatically approved. Alternatively, it can be sent to the administrator to approve or reject the request. If the request is approved, the correspondent changes are applied.

Two days before the end date of the project, the team administrator is notified, and he or she can request to modify the end date of the project. When the end date and time arrive, or if the project is canceled by the user, TPM saves the VMs images (if requested by the user) and starts to shut down the VMs; in this stage, the project status is "Decommissioning". After the VMs are shut down, they are erased from the system and the resources are freed, then the project status is "Decommissioned" and the team administrator is notified that the project or the service ended.

11.6   Experimental Results

During the implementation, some scenarios were tested to see how fast the service request will be offered to the user. The executed tests were based on deploying a number of virtual machines, using the service described in the above paragraphs.

The service request used was for projects that required immediate deployment; the deployment process can be split (from the point of view of deployment duration) into the following stages:

- *Querying and preparing the infrastructure for deployment* (querying resources, writing infrastructure status in the database, communicating with hypervisors and vSphere, etc.)—this process takes about 10 min for each project and is not varying too much with the number of VMs in the project. For example, for a project with 10 VMs, the time is about 9.41 min, and for a project with 100 VMs the time is about 10.3 min.
- *Cloning the VMs*—this process is executed on the blade servers that access the DS3400 storage; the cloning process is done in parallel, using 8 threads (8 VMs are cloned simultaneously from the same source file). The storage uses 3 expansions, offering a total raw space of 29 TB of data, configured in three RAID5 arrays. The measured writing speed is about 300 MB/sec. The time to clone 10 VMs is about 14.25 min, and for 100 VMs is about 55.51 min when the size of the VM HDD is 10 GB.
- *VM hardware reconfiguration*—this stage is executed in order to reconfigure the VM hardware (number of CPU, RAM, HDD expansion, adding network adapters).

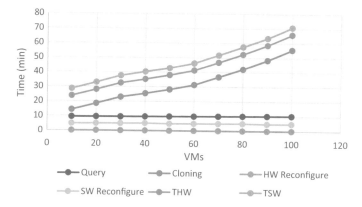

Fig. 11.8 Deployment time

It is executed very fast by the hypervisor, in 6 s, this time being added to the total deployment time of the project.

- *VM software reconfiguration*—this stage is optional and adds more time to the project, depending on the size of the SN-MaaS tools and the installation procedure. In our case, the software had about 50 MB and the installation takes about 5 min. Because the installation of the additional software is not limited to 8 threads, this stage will not scale with the number of VMs into the project.

The results of the experiments are presented in Fig. 11.8. The tests were executed with projects starting from 10 VMs, and then progressively adding sets of 10 VMs, up to the last project, which had 100 VMs. All VMs were created based on a Windows 7 32 bit template, with HDD of 10 GB, reconfigured to 60 GB after cloning. The measured deployment time is presented (in minutes) for the query stage, cloning stage, hardware reconfiguration, software reconfiguration (additional software installation); the figure also illustrates the total deployment time without additional software installation (THW), and the total deployment time with additional software installation for SN-MaaS.

11.7   Conclusion

The chapter presented a solution for Sensor Network Modeling as a Service, used by the master's students at University Politehnica of Bucharest for creating specific models and then programming various model interpreters. It was implemented in cloud based on Infrastructure as a Service and offering an educational service under the form of virtual machines provisioned on demand. The system provides access to virtual machines preinstalled with the operating system and the sensor network modeling environment, to be immediately used by students and teachers for educational or research purposes.

Being an on-demand service, the responsiveness and the deployment speed are very important; to measure the efficiency, multiple deployment tests were executed, creating projects that included 10, 20, 30 up to 100 virtual machines. The system deployed the projects with an appropriate speed, i.e., about 55 min for a project with 100VMs. The tests proved that the deployment performance is better when the SN-MaaS tools are preinstalled and contained in the VM template, with respect to the situation when the SN-MaaS tools are installed after the VM cloning. The system offers the students the possibility to work with SN-MaaS tools not only from the university laboratories but also from home, allowing them to gather better knowledge and skills in the fields of sensor networks and modeling languages.

Acknowledgments This work was supported by a grant of the Romanian National Authority for Scientific Research and Innovation, CNCS/CCCDI—UEFISCDI, project number PN-III-P2-2.1-PED-2016–1336, within PNCDI III.

References

1. Ahmed K, Gregory M (2011) Integrating wireless sensor networks with cloud computing. In: 2011 seventh international conference on mobile ad-hoc and sensor networks. Beijing, pp 364–366
2. Alamri A, Ansari WS, Hassan MM, Hossain MS, Alelaiwi A, Hossain MA (2013) A survey on sensor-cloud: architecture, applications, and approaches. Int J Distrib Sens Netw 2013, 917923, 18
3. Alnabhan N, Al-Aboody N, Al-Rawishidy H (2017) Adaptive wireless sensor network and cloud-based approaches for emergency navigation. In: Demonstration at the 42nd ieee conference on Local Computer Networks (LCN). Singapore. https://www.ieeelcn.org/prior/LCN42/lcn42demos/1570388472.pdf
4. Applying the cloud in education. An innovative approach to IT (2012) IBM Global Technology Services White Paper
5. Bouyer A, Arasteh B (2014) The necessity of using cloud computing in educational system. Procedia—Social Behavioral Sci 143:581–585
6. Cheekiralla S, Engels DW (2005) A functional taxonomy of wireless sensor network devices, BroadNets 2005. In: The 2nd international conference on broadband networks. Boston, MA, pp 949–956
7. Chung WY, Yu PS, Huang CJ (2013) Cloud computing system based on wireless sensor network. In: Proceedings of the 2013 federated conference on computer science and information systems, pp 877–880
8. Darmawan B, Siglen J, Lundgren L, Catterall R (2008) IBM tivoli usage and accounting manager V7.1 handbook. International Technical Support Organization, ISBN-10: 0738485640, ISBN-13: 9780738485645
9. Dash SK, Mohapatra S, Pattnaik PK (2010) A survey on applications of wireless sensor network using cloud computing. Int J Comput Sci Emerging Technol 1(4):50–55
10. Delicato FC, Pires PF, Pirmez L, Batista T (2010) Wireless sensor networks as a service. In: 17th IEEE international conference and workshops on engineering of computer-based systems, pp 410–417
11. DiMauro G, Scalera M, Visaggio G (2016) The educational cloud, problems and perspectives. In: Proceedings of the 20th world multi-conference on systemics, cybernetics and informatics (WMSCI 2016), pp 34–40

12. Flammini A, Sisinni E (2014) Wireless sensor networking in the internet of things and cloud computing era, EUROSENSORS 2014, the XXVIII edition of the conference series. Procedia Eng 87: 672–679
13. Fogel R (2010) The education cloud: delivering education as a service. WHITE PAPER, Intel® World Ahead
14. Fokum TD, Frost VS, Mani P, Minden GJ, Evans JB, Muralidharan S (2008) A taxonomy of sensor network architectures, Technical Report ITTC-FY2009-TR-41420–08, The University of Kansas
15. GME (2017) Generic Modeling Environment. http://w3.isis.vanderbilt.edu/Projects/gme/. Accessed 14 Oct 2017
16. Gucer V, Altaf N, Anderson ED, Boldt DG, Choilawala M, Escobar I, Godfrey SA, Sokal MM, Walker C (2008) IBM Tivoli monitoring: implementation and performance optimization for large scale environments, international technical support organization, ISBN-10: 0738488593, ISBN-13: 9780738488592
17. Hadzhikoleva S, Hadzhikolev E, Cheresharov S, Yovkov L (2018) Towards building cloud education networks. TEM J 7(1):219–224
18. Hang L, Jin W, Yoon H, Hong YG, Kim DH (2018) Design and implementation of a sensor-cloud platform for physical sensor management on CoT environments. Electronics 7: 140
19. Huché T, Koohi B, Lam TV, Reynolds P, Swehla SM, Wain J (2012) Cloud computing infrastructure on IBM power systems, getting started with ISDM, international technical support organization, ISBN-10: 0738436674, ISBN-13: 9780738436678
20. Ionita AD, Anton FD, Olteanu A (2018) Sensor network modeling as a service. In: Proceedings of the 8th international conference on cloud computing and services science (CLOSER 2018), pp 346–353
21. Ionita AD, Gurau A (2015) Visual modeling environment for sensor networks. In: ISEF 2015—XVII international symposium on electromagnetic fields in mechatronics, electrical and electronic engineering. Valencia, Spain
22. Kannadasan R, Prabakaran N, Boominathan P, Krishnamoorthy A, Naresh K, Sivashanmugam G (2018) High performance parallel computing with cloud technologies. Procedia Comput Sci 132:518–524
23. King TS (2015) Reviews of cloud computing for education: services and benefits. PEOPLE: Int J Soc Sci 1(1): 1299–1305
24. Lemos A, Moleiro R, Ottaviano P, Rada F, Widomski M, Braswell B (2012) IBM CloudBurst on system x, international technical support organization, ISBN-10: 0738436526, ISBN-13: 9780738436524
25. Maresova P, Kacetla J (2015) Cloud computing in the public sector—case study in educational institution. In: 4th world conference on educational technology researches, WCETR2014, Procedia—Social and Behavioral Sciences, vol 182, pp 341–348
26. Morariu O, Borangiu Th (2012) Resource monitoring in cloud platforms with tivoli service automation management. IFAC Proc Volumes 45(6):1862–1868
27. Olivieri A, Pintus A, Mallozzi A, Santucci C, Allocca D, Salustri F, Santoro G, Maniscalco G, Shah G, Lau J, Balestrazzi L, von der Planitz ME, Bove R, Mariottini V, Gucer V, Calafiore V, Leung WL (2009) IBM Tivoli provisioning manager V7.1.1: deployment and IBM service management integration guide, international technical support organization
28. OpenGIS (Open Geospatial Consortium) (2007) Sensor Model Language (SensorML) implementation specification, M. Botts Ed., Version: 1.0.0
29. Papazoglou MP (2018) Smart connected digital factories. Unleashing the power of industry 4.0 and the industrial internet. In: proceedings of the 8'th international conference on cloud computing and services science (CLOSER 2018), pp 5–16
30. Rădulescu ŞA (2014) A perspective on e-learning and cloud computing. Procedia—Soc Behavior Sci 141:1084–1088
31. Sheng X, Xiao X, Tang J, Xue G (2012) Sensing as a service: challenges, solutions and future directions. IEEE Sens J 13(10):3733–3741

32. Trends in Cloud Computing in Higher Education (2015) ViON White Paper, eCampus-News, June 29. https://www.ecampusnews.com/whitepapers/trends-in-cloud-computing-in-higher-education/
33. White RM (1987) A sensor classification scheme. IEEE Trans Ultrason, Ferroelectr, Freq Control UFFC-34(2): 124–126

Chapter 12
Methods and Techniques for Automatic Identification System Data Reduction

Claudia Ifrim, Manolis Wallace, Vassilis Poulopoulos, and Andriana Mourti

Abstract Extensive use of sensors and system automations tend to become a trend in our everyday life. In this manner, maritime traffic is monitored by advanced sensor systems, one of which is the Automatic Identification System, called AIS. The use of AIS systems in shipping is extensive and large numbers of data are produced; thus the task of analyzing this information becomes a laborious procedure. In this paper, we analyze AIS data sets in order to propose methods and techniques for reducing the data without losing important information in order to produce sets of data that can be easily processed in real time. Furthermore, the deriving data are visualized so as to present that data reduction does not affect the visualized data which is a very common application of AIS data.

12.1 Introduction

Nowadays, sensors and intelligent systems are widely used and extensive research and development activities are performed in this domain. The outcomes of the research that is done is available to our everyday lives, even without being able to realize the existence or the complexity of the systems behind daily tasks. An important intelligent sensor system that is used primarily for identifying and locating vessels is Automatic Identification System (AIS). The Automatic Identification

C. Ifrim
University Politehnica of Bucharest, Bucharest, Romania
e-mail: claudia.ifrim@hpc.pub.ro

M. Wallace (✉) · V. Poulopoulos · A. Mourti
Knowledge and Uncertainty Research Laboratory, University of Peloponnese, Tripoli, Greece
e-mail: wallace@uop.gr

V. Poulopoulos
e-mail: vacilos@uop.gr

A. Mourti
e-mail: andrianamourti@uop.gr

© Springer Nature Switzerland AG 2021
F. Pop and G. Neagu (eds.), *Big Data Platforms and Applications*,
Computer Communications and Networks,
https://doi.org/10.1007/978-3-030-38836-2_12

System (AIS) is an automated tracking system used on ships and by vessel traffic services (VTS) that broadcasts in an interval of seconds information, such as unique identification of the ship, position, course, speed, and navigation status, to other nearby ships, AIS base stations and satellites [1].

Acting as an enhanced two-way positioning system, it is able to assist the tracking and monitoring of vessel movements to watch-standing officers, by providing a number of useful metadata. On board, it is integrated with other navigation aids, such as:

- Global Positioning System (GPS);
- Radio Detection And Ranging (RADAR);
- Electronic Chart Display Information System (ECDIS);
- Voyage Data Recorder (VDR);
- Automatic Radar Plotting Aid (ARPA).

As it is obvious from several different options, the number of positioning and navigation aiding systems is large and, unfortunately, do not utilize a unique standard. Such a policy would offer a global solution for actively managing the increasing traffic and provide solutions for efficient resource planning. This means that the analysis of maritime traffic data would lead to the analysis of a very large number of heterogeneous data. In specific, AIS data received by base stations or satellites is recorded to stations for extended use, either real-time or future use. According to the protocol that defines the generation of AIS data, their number is vast and can easily overload a storage system in a short period of time; therefore, leading to an increase of the amount of time needed for data processing and information retrieval.

This paper presents methods and technologies applied on the nature of AIS data sets in order to achieve storage of much lower amounts of data without losing the important information about monitored vessels. The actual size of data produced for a number of vessels can easily become unmanageable, leading to an increase in the needs of resources for the applications that use this data for historic or real-time purposes. We believe that lossless data reduction can decrease the response time of such applications and provide better quality in the combination of results.

The rest of the paper is organized as follows: The next section presents projects and research that rely on analysis of AIS data. Section 12.3 contains an overview of AIS technology, what types of errors can be detected on AIS data, and how they can be corrected, as well as references to the existing applications; Sect. 12.4 presents methods and techniques for reducing the amount of AIS data without losing important information; in Sect. 12.5, we present the results of applying our techniques on AIS data sets and we finalize with Sect. 12.6 discussing the results of our procedure and future enhancements on the proposed methodologies.

12.2 Related Work

A number of research fields is related on maritime traffic each one for different purposes. As the AIS data is useful for providing details about the ship, its cargo and its route information, they are utilized for their quality of information.

As part of the project "Emissions from shipping in the Arctic" in [2] they clearly state that the availability of satellite based AIS data for ship tracking makes it possible to set up detailed fleet specific emission inventories in a high temporal and spatial resolution for the Arctic and carry out dispersion calculations with enhanced precision. Similar to this, a project that analyzed emissions in Asia was conducted and presented in [3].

An unsupervised and incremental learning approach to the extraction of maritime movement patterns is presented in [4] to convert from raw data to information supporting decisions. This is useful, for example, in counter piracy applications to identify risk areas associated with the joint predicted presence of white shipping density (e.g., commercial merchant traffic) and Pirates Action Groups (PAG) [5].

The possibility of using AIS data for fisheries research and the provision of an analysis about the level of uptake of the AIS by the EU fishing fleet is explored in [6]. The specific work is part of a more long term objective of producing a high resolution map of fishing effort for Europe using AIS. A similar procedure for improving the detection of fishing patterns from Satellite AIS data using Data Mining and Machine Learning is presented in [7].

From the aforementioned, it is clear that the applications of AIS data analysis can be multidimensional. It includes from simple fleet management or monitoring, environmental impact of vessels to more complex and important tasks that include real-time collision avoidance systems. As the regulations are such that AIS has to be installed to every vessel, it is a technology that will formulate the future of data analysis considering maritime traffic.

12.3 AIS Technology

The Automatic Identification System (AIS) [8] is an automated, autonomous tracking system which is extensively used in the maritime world for the exchange of navigational informational between AIS-equipped terminals [9]. As its definition implies, it is a two-way navigation and metadata exchange system designed to be applied on vessels. In order to operate as a two-way system, it consists of AIS receivers and transmitters on board, satellite, and ashore; which enable the electronic exchange between AIS stations. In fact, all passengers' vessels and all commercial vessels over 299 Gross Tonnage (GT) that travel internationally are required to have an AIS transponder aboard. Two types of transponders exist, Class A, which should be used generally by all vessels, while smaller vessels can use Class B transponders which provide fewer data and have smaller ranges of transmission (5–10 min) [1].

The AIS format uses Time-division multiple access (TDMA) which is a channel access method for shared-medium networks. It allows several users to share the same frequency channel by dividing the signal into different time slots [1]. Each user is assigned a time slot during which the transmission is possible, making the procedure able to be performed at the same time to a single station from a large number of transmitters. On the other side, the time slots are just 4,500 per minute; assuming each vessel requires a time slot for its transmission, this numbers is equal to the maximum capacity of a station. In case of larger number of transmitters, interference between the signals will occur, and as a result a number of errors will be produced to the recorded data. In order to overcome such a problem, a grid of receivers is used in cases that is expected to have transmissions of large number of data from multiple vessels at the same time and place. Apparently, due to the nature of maritime traffic, this is expected to occur in specific times and places which makes it possible to foresee and anticipate the problem. In more detail, multiple base stations are installed in high traffic areas, ports and channels; all of them aggregate their input into a single stream of data while in parallel they correct multiple instances of the same information (duplicates) or errors that may occur.

Technically, AIS data is defined as ASCII[1] packets as byte stream using the NMEA 0183 [10] or NMEA 2000 [11] data formats. As a broadcasting transmission protocol, it has an indicator for packets concerning other ships ("!AIVDM") and packets concerning current ship ("!AIVDO"). The standard about the so-called AIVDM/AIVDO messages is ITU1371 [12], which was expanded and clarified by ITU-R [13]. The ASCII format for AIVDM/AIVDO representations of AIS radio messages have been set by IEC-PAS [14] and a common AIVDM byte stream could be like the following:

!AIVDM,1,1,B,177KQJ5000G?tO'K > RA1wUbN0TKH,0*5C

The following Table 12.1 presents the definition of each part of the message.

The *-separated suffix (*5C) is the NMEA 0183 checksum for the sentence, preceded by "*". It is computed on the entire sentence including the "AIVDM" excluding the "!"

As the AIS protocol is used for defining many different parameters under many different circumstances, there are numerous different types of messages. Twenty-seven different message types exist, most of which are used to report position. Regulations exist in order to define the frequency of transmitting by type of message. As such, *safety* messages must be sent as required, *long range* messages have to be sent every 30 min, *static* messages have to be sent every 6 min or when data is amended upon request, and finally, the most important *dynamic* messages have very detailed rules that must be followed in order to achieve the scope of the AIS protocol usage [15]. The following Tables 12.2 and 12.3 present the dynamic message frequency regulations by type:

From the definition of the regulations, it is obvious that messages can occur as often as every 2 s per vessel. The regulation and definition of the protocol is such that

---

[1] ASCII character set—reference definition: https://tools.ietf.org/html/rfc2046.

Table 12.1 Definition of AIS data example

Field	In example	Definition
#1	!AIVDM	Identifies this as an AIVDM packet
#2	1	Is the count of fragments in the message. The size of a sentence is limited to an 82-character maximum, so it sometimes has to be split over multiple sentences
#3	1	Is the fragment number of this sentence, one based. So a message with fragment count of 1 and fragment number of 1 is a complete message
#4	Empty	Is a sequential message ID for multi-sentence messages
#5	B	Is a radio channel code. AIS uses the high side of the duplex from two VHF radio channels: AIS Channel A is 161.975 Mhz (87B); AIS Channel B is 162.025 Mhz (88B). Codes 1 and 2 may also be encountered instead of A or B
#6	177KQJ5000G?tO'K> RA1wUbN0TKH	Is the data payload
#7	0	Is the number of fill bits requires to pad the data payload to a 6 bit boundary, ranging from 0 to 5

Table 12.2 Dynamic Class A

Ships dynamic conditions	Not changing course	Changing course
At anchor or moored and moving less than 3 knots	3 min	3 min
At anchor or moored and moving faster than 3 knots	10 s	10 s
0 to 14 knots	10 s	3 1/3 s
14 to 23 knots	6 s	2 s
Over 23 knots	2 s	2 s

it acts as the medium to achieve a number of important issues that exist in maritime traffic. In short, the transmission of the messages can support:

- Collision avoidance
- Vessel traffic services
- Aids to navigation
- Search and Rescue
- Maritime security
- Cargo tracking
- Fleet tracking
- Fishing fleet monitoring.

Table 12.3 Equipment other than Class A shipborne mobile

Platform's condition	Nominal reporting interval
Class B "SO" shipborne mobile equipment not moving faster than 2 knots	3 min
Class B "SO" shipborne mobile equipment moving 2–14 knots	30 s
Class B "SO" shipborne mobile equipment moving 14–23 knots	15 s
Class B "SO" shipborne mobile equipment moving > 23 knots	5 s
Class B "CS" shipborne mobile equipment not moving faster than 2 knots	3 min
Class B "CS" shipborne mobile equipment moving faster than 2 knots	30 s
Search and rescue aircraft (airborne mobile equipment)	10 s
Aids to navigation	3 min
AIS base station	10 s

Each of the aforementioned usages of the AIS data may have different prerequisites. In our algorithmic analysis, we focus both on the kind of information that is needed for real-time application, such as collision avoidance, as well as data analysis application, which can occur in research fishing fleet analysis.

12.4 Algorithm Analysis

From the analysis and definition of the protocol, we observe that a large number of data is produced in a unit of time, in which we should be able to apply algorithms and procedures either to reduce the data themselves or to predict data. By analyzing the types of messages existing (as presented in the following list), we put the focus on position messages which are the most frequent (messages 1, 2, 3, 18 and derivatives 5, 19, and 24). The following Table 12.4 displays the AIS messages.

Table 12.4 AIS message analysis

Message	Analysis
1, 2, 3	Position reports
4[a]	Base station report
5	Ship static and voyage related data
18	Standard Class B equipment position report
19	Extended Class B equipment position report
24	Static data report

[a]Message 4 will be treated just to display the general geographic distribution of base stations

Important information that helps us recognize and analyze the messages that are transmitted are the number of ships or equipment of the transmitter, a number that is unique; an identification of the message transmitted is helpful for distinguishing messages as well as a repeat indicator that was designed to be used for repeating messages over obstacles by relay devices.

We are mainly focusing on messages 1, 2, and 3 (position reports for class A) as they contain navigational information that include longitude and latitude, time-stamp, heading, speed, ship's navigation status (under power, at anchor, etc.), which are the messages on which we will apply our algorithmic procedures.

A surface observation of the data leads to more than 2 million records in a small but crowded area in the time of a month. In parallel, more and more vessels install AIS devices on board leading to an exponential increase of the data in the period of time. Consequently, it is expected that data will only increase and the effort of the applications analyzing data will be affected. As a matter of fact, it is inevitable that reducing the amount of data without losing the valuable information should be a procedure that is essential.

A first procedure of our algorithm is the identification of the parameters that are included in the majority of messages and could be candidate data to be compressed. As already mentioned in the previous paragraphs, messages 1, 2, and 3 containing navigational information will be processed.

From the data observation, as well as the nature of the protocol, we make some important assumptions in order to examine if there are situations under which we should be able to reduce data. We conclude the following four situations which are candidates to help us reduce the data.

Repeated data. This is a very common situation that can occur when we receive data from "stopped" vessels. A vessel can be in multiple different conditions in order to be stopped (e.g., vessel in port, or standing for refueling). In that case, we can apply a simple algorithm that rejects new data (e.g., when latitude and longitude remain unchanged) and update the time-stamp of the latest stream.

Easily calculated data. This approach has the philosophy of Video compression protocols. What remains unchanged or can be easily calculated (e.g., standard route) can be omitted as information. *Huge number of data.* For specific occasions where the vessels' speed is too low, despite the fact that the protocol itself predicts lower rates of transmission, we can enhance the procedure with additional algorithmic procedures in order to further reduce the data. *Custom data provision.* As a matter of fact, we are able to provide an API through which we perform custom queries on the data in order to return as results portions of the data and not the complete data set. In this manner, we should be able to perform all our techniques on the data that are demanded and only the part of data that is useful is provided as an answer.

In this manner, we utilize a library in order to be able to decode AIS streams from a specific AIS data set. This is useful as we will be able to have the encoded byte stream as "human readable" information on which we are able to perform algorithmic

actions. It is mC++[2] decoder for Automatic Identification System for tracking ships
and decoding maritime information.

By analyzing our AIS data set, we can easily observe that the records count can be
2 million or more per month. Considering the increased number of vessels that install
AIS devices on board, the number of records that will be stored can only increase
and the only solution that we have is to propose a reduction technique that could be
applied on AIS data records in order to reduce its size without losing any valuable
information and to be able to optimally store and query historic AIS data.

We identified that navigational messages are the ones that will be processed in
order to reduce either their number or their information. A vessel broadcasts such
messages within specific time intervals that are based on its speed and course. By
analyzing several different messages transmitted by a single vessel on voyage, we
make the following assumptions:

- Attributes frequently changing over time: *latitude, longitude, time-stamp*
- Attributes rarely changing over time: *speed, heading.*

From the aforementioned, we extract the following information. When we are
aware of a starting location, the heading and the speed we may locate the final
location if we are aware of the starting and final time-stamp. This assumption helps
us remove a large number of the frequent messages and replace them with a start
and final location and the period of time. This also means that a vessel is traveling in
straight lines, and thus we should be able to have enough information to reproduce
the vessel's path with a starting and ending point. Under other conditions, if the speed
remains under 0.1 knots, we should be assured that a vessel is not moving or that the
change of its location should be imperceptible.

12.4.1 Analyzing the Data Set

From the analysis described, it is more than obvious that each vessel's records must
be treated separately. If we obtain a data set from records deriving from a base station,
that are the transmitted data from the vessels in proximity, then we should be able to
recognize each vessel's transmissions. At first, we extract all the unique MMSI values
in order to obtain the number of vessels that transmitted AIS data. For each of the
extracted unique identifier, we export the byte streams transmitted in chronological
order. We are utilizing the technique of "windows of information" a procedure similar
to the way video compression is implemented. We assume that we have *i-records* that
are records which include the complete set of data for all the parameters, *d-records*
that are records which include only data for parameters that were differentiated and
finally *p-records* that are records whose data for the parameters can be fully predicted.
As in the video compression procedure some "key records" contain all the data while
every other record's data are reduced significantly. Depending on the cruising point

---

[2] Libais C++ AIS data decoder: https://github.com/schwehr/libais/tree/master/src/libais.

of a vessel, it is expected to have a small number of *i-records* whenever a ship is en-route that include a large number of d or p-records, while in the situation of ship maneuvering more i-records are essentials while predictions or differentiation cannot be expected in the transmitted data.

12.5 Experimental Evaluation

The AIS data set that we used for our experiments contains information retrieved from the area of the Black Sea. It includes a number of 136,008,000 records. At first, we perform data correction, as well as data cleansing, a data preprocessing procedure in which we remove incorrect information. After this procedure, we apply the data reduction techniques and we visualize the results before and after the reduction in order to present the differentiation of the procedure.

12.5.1 Analyzed Data

The initial data analysis of the information leads to the creation of two different database tables. A PostgreSQL database with PostGIS extension is used to store the information within the messages. The information recorded was the decoded information and thus the two tables include:

- *static information.* Includes all the data that are related to the physical information of a vessel (type of vessel, length, year of construction, etc.)
- dynamic information (latitude, longitude, speed, etc.).

Despite the fact that the static information can provide us with a large number of qualitative data, which could be, for example, the maximum speed of the vessel and can be helpful for determining data anomalies, we analyze the dynamic data as they are much larger in number and are the data that can be used for applying data reduction techniques.

12.5.2 Data Reduction Applied on AIS Data Set

As mentioned in the previous paragraphs, some information is eliminated as an initial cleanup procedure. This includes searching for the following malformed data and removing them:

- Coordinates greater that 180, −180 latitude, and 90, −90 longitude
- The 0, 0 location.

Table 12.5 Break down of record types produced by algorithm application

Initial	Information records	Difference records	predicted records
568,934	96,719	223,488	248,727
	17%	39%	44%

This procedure removes almost 20% of the records in the data set, leading to the assumption that a large number of the transmitted data contain errors, that cannot possibly be corrected. The experimental evaluation is focused around a specific open sea area between Constanta and Sevastopol, which is crowded with AIS data transmissions (point: 43.70, 26.60). As a matter of fact from the starting set of data, after removal of the false and by applying a proximity query, we are able to obtain a number of 568,934 records.

The experimental evaluation consists of algorithm application on these records to obtain the final number of records that occur, indicating the compression that can be done on data. The application of the algorithm implies that it is possible to receive a number of almost 17% (96,719) of the data as information positions, and thus we need to keep the complete set of data. In order to be assured that we will have a number of detailed information records, it is essential to record one i-record at least every 20 min. These data cannot be compressed as they are essential in order to calculate the differentiated data (d-records) as well as the predicted data (p-records).

According to the algorithm, whenever we have slight changes to the speed then we are able to remove any other information and store the speed difference only. This is a d-record. In case of unchanged speed or heading, then a p-record is produced which is actually a virtual records as we assume that data for the specific time-stamp can be predicted. As the vessels tend to have standard speed and straight paths when being in open sea, then we are able to have large number of prediction and an amount of differentiated data. The following Table 12.5 presents the number of records per record type.

By analyzing the size of data, we shall be able to compute the level of compression. The information data include a number of data that need to be stored in the database in order to recognize and analyze AIS data. On the other hand, the data stream that is sent when a transmission exists could be a small number of bytes compared to the data stored in a database record. We will make a comparison with both the database info and the byte data stream. The data stream as defined by NMEA 0183 protocol is limited to a maximum of 82 characters which can be equivalent to the number of bytes needed to store the "sentence". On the other hand, a database human readable record of the data is more than 240 byte long, as a number of metadata is stored. The total number of 568,934 records would require 46MB of AIS data sentences or 136MB of data in the database. The number is approximate based on the maximum data. Our algorithm can possibly remove at all almost 45% of the messages, which are messages that are able to be predicted by combination of the information records and the differentiations records. Thus, the space required is only 55% of the initial. In

Table 12.6 Initial records versus reduced records

Initial number of records	Unique MMSI	Speed/heading difference	Final number of records
752,552	458	Less than 0.1/less than 5 degrees	248,743
752,552	458	Less than 0.15/less than 3.5	204,338
752,552	458	Less than 0.2/less than 1	202,248

parallel, when we need to store only the differentiation, then we put the focus on the speed parameter. In almost 40% of the cases, we can keep only the speed difference related to the information record of reference. Storing this data can be as large as 3 bytes in extreme occasions. This means that in 40% of the cases we can have a compression of 96% compared to the data stream and 98% compression compared to the data stored in the database records. Applying the compressions in the initial data, we conclude that we have a compression of 37% in the case of the byte stream and 38% in the case of records in the database. The total compression achieved by application of the algorithm is considered to be almost 80%. This compression can be easily achieved in areas that are open sea points of sea traffic where vessels are expected to have standard route and speed.

Our algorithm treats differently areas where large amount of diverse traffic occurs, which can be the are of the port. In this case, we examine the Constanta port. The initial set for the area has 752,552 records occurred from 458 different vessels. For this set, we followed the algorithm using different parameters for speed and heading of the vessels. The algorithm implies that when the speed or heading values change slightly then we omit the records considering them as ones that can be predicted or just be thrown away. As an example, we can consider that a speed of 0.1 knots means that a vessel will move around 150 m within an hour. At this amount of time, the system will certainly have a number of updates, thus making it possible to have a working data set without the need of this data (Table 12.6).

In this case, we observe that by completely omitting the records that have low speed or slight difference in speed heading, we can achieve a compression of almost 75%. The following section presents the visualization of the real and differentiated data set on a map. By viewing the initial versus reduced amount of data, it is possible to realize that our data reduction algorithm does not seem to cause any real change in the data, and especially of what is needed by AIS and is the definition of data, speed, heading, and location of vessels. From the visualization, it is furthermore possible to apply higher margins in speed and heading, and have a much large compression (more than 85%).

12.5.3 Data Visualization

In order to understand the differences that occur in the initial data set compared to the data set produced after application of our algorithm, we present a set of visualizations. By viewing the representation of the data on the map, we actually depict each record of the database as a dot on the map. The first set of visualizations present how the initial data set is presented on the map. In Fig. 12.1, we can see the row data presented on the map, while Fig. 12.2 presents the same data organized by MMSI, noting that MMSI is unique for each different vessel. Using a visual comparation we can easily outline the impact of our solution on data size and data quality. In this section, we present the visualization of the initial data set, the visualization of the reduced data set using different parameters for the speed difference, and in the last image we will have a representation of the initial data set compared with the reduced data set resulted after we applied our reduction algorithm.

Finally, Fig. 12.3 presents the density of the transmitted messages related to their position on the map. It is obvious that there is a huge number of transmitted data and the density—deriving from the port of Constanta—is large, as it is expected to be in huge sea traffic areas.

After the initial figures, we proceed with visualization of the data that are generated after the application of the algorithm on the initial data set. We will focus on the port of Constanta as it was the area that was analyzed in the experimental evaluation producing data reduced by 75% compared to the initial.

Fig. 12.1 Initial data set—routes visualization

Fig. 12.2 Initial data set—routes visualization by MMSI

Fig. 12.3 Initial data set—density map

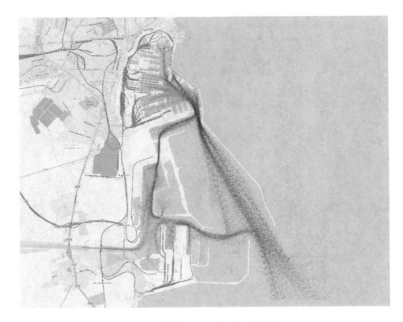

Fig. 12.4 Compressed data set—speed difference less than 0.1 knots

It is obvious from Fig. 12.4 that for the simplest application of our algorithm, which is expected to have the worst compression rate we observe that it seems that the visualization remains almost the same. As the density map depicts in Fig. 12.3, the data are so dense in the specific area—expected to be the case of any busy port—that omitting a large number of data does not seem to affect the data set.

We furthermore performed the visualization in extreme values of the parameters of the algorithm, which is altering the speed parameter limit to 0.5 knots and 1 knot. *We should note that with 1 knot speed a vessel can move 1.8 km within an hour.* It is obvious from Figs. 12.5 and 12.6 that the data loss is such that the result can be acceptable as long as we are able to predict and reproduce the omitted data, a case that is possible according to our algorithm definition.

Finally, we calculate the aggregated results of comparison between the initial data set and the compressed data. In this occasion, we decided to compare the differences that occur in situations of parameter limit for speed being below 0.2 knots.

It is obvious from Fig. 12.7 that within the area of the port, where speeds tend to be lower the data set is reduced significantly. In the area outside the port, the reduction is less with the application of the speed limit parameter, though in this area, the compression of data occurs from the algorithm part that creates differentiation records and prediction records.

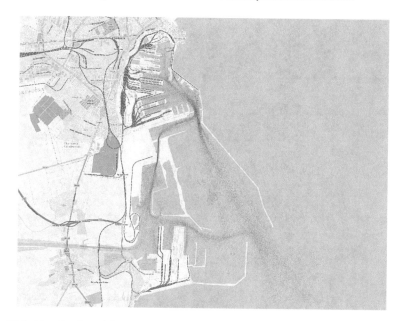

Fig. 12.5 Visualization of the reduced data set for extreme cases|speed difference less than 0.5 knots

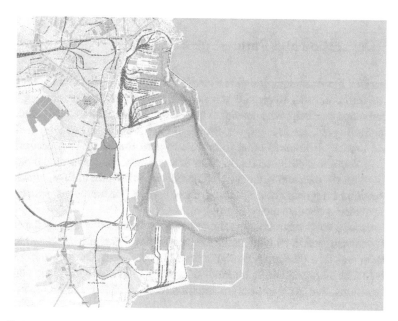

Fig. 12.6 Visualization of the reduced data set for extreme cases|speed difference less than 0.5 knots

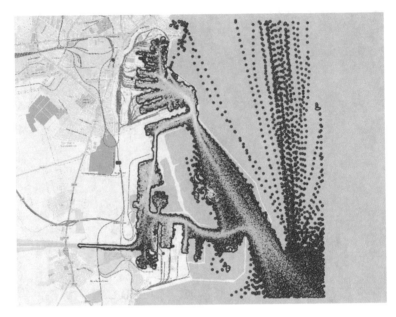

Fig. 12.7 Aggregated results—initial data set versus reduced data set (speed difference less than 0.2 knots)

12.6 Conclusion and Future Work

We described the Automatic Identification System (AIS) a system that is utilized in maritime traffic in order to provide a set of functionality including, among others, procedures that can help even special occasions including, collision avoidance and fleet monitoring. In this manner, we discussed the necessity of having real-time or near real-time application in order to be able to perform the set of procedures that the AIS protocol was designed for. In order to serve its scope, the protocol has strict regulations on the data and amount of data produced in the unit of time, which leads to generation of huge databases within a short period; thus making the analysis of the information a big data analysis problem.

In this scope, we presented a novel approach for significantly reducing the amount of data produced by AIS without losing the information that could be needed in order to perform real-time data analysis and actions required by it. Our algorithm is able to analyze data and create different kinds of records similarly to the video compression algorithms; that is creation of information records, differentiation records, and prediction records.

The experimental evaluation was performed on a large data set produced in a dense traffic area and more precisely in the port of Constanta located Romania (Black Sea). We divided our algorithm into two main occasions: places with very high density of data due to low speeds (usually ports) and areas with high density of data due to standard routs. We presented how we can reduce the produced data in both occasions without losing the information. Our experiments prove that we can perform compression of at least 75% in both types of areas. We presented visualization of the initial and derived data in order to prove the information persistence after the compression.

As a future work, we plan to create a real-time service for analyzing in real time the data produced by AIS and provide a near real-time API that will be able to provide compressed AIS data. We will furthermore elaborate on the parameters of the algorithm in order to achieve more efficient levels of compression.

Acknowledgements This work has been partially supported by COST Action IC1302: Semantic keyword-based search on structured data sources (KEYSTONE); we particularly acknowledge the support of the grant COST-STSM-IC1302-36978: "Curating Data Analysis Workflows for Better Workflow Discovery".

References

1. What is the automatic identification system (AIS)? https://help.marinetraffic.com/hc/en-us/articles/204581828-What-is-the-Automatic-Identification-System-AIS
2. Winther M, Christensen JH, Plejdrup MS, Ravn ES, Eriksson ÓF, Kristensen HO (2014) Emission inventories for ships in the arctic based on satellite sampled AIS data. Atmos Environ 91:1–14, ISSN 1352-2310
3. Chen D, Wang X, Li Y, Lang J, Zhou Y, Guo X, Zhao Y (2017) High-spatiotemporal-resolution ship emission inventory of China based on AIS data in 2014. Sci Total Environ 609
4. Pallotta G, Vespe M, Bryan K (2013) Vessel pattern knowledge discovery from AIS data: a framework for anomaly detection and route prediction. Entropy 15(6):2218–2245
5. Hansen J, Jacobs G, Hsu L, Dykes J, Dastugue J, Allard R, Barron C, Lalejini D, Abramson M, Russell S et al (2011) Information domination: dynamically coupling METOC and INTEL for improved guidance for piracy interdiction. NRL Rev 2011:110–119
6. Natale F, Gibin M, Alessandrini A, Vespe M, Paulrud A (2015) Mapping fishing effort through AIS data. PLoS ONE 10(6):e0130746
7. de Souza EN, Boerder K, Matwin S, Worm B (2016) Correction: improving fishing pattern detection from satellite AIS using data mining and machine learning. PLOS ONE 11(9)
8. Greene M (1993) Radio frequency automatic identification system. U.S. Patent No. 5204681
9. Automatic identification system. https://en.wikipedia.org/wiki/Automatic_identification_system
10. Definition of the NMEA 0183 Standard. http://www.nmea.org/content/nmea_standards/nmea_0183_v_410.asp
11. Definition of the NMEA 2000 Standard. http://www.nmea.org/content/nmea_standards/nmea_2000_ed3_10.asp
12. ITU Recommendation M.1371, Technical characteristics for a universal shipborne automatic identification system using time division multiple access [ITU1371]
13. IALA Technical Clarifications on Recommendation ITU-R M.1371-1
14. IEC-PAS 61162-100, Maritime navigation and radio communication equipment and systems [IEC-PAS]
15. Harati-Mokhtari A et al (2007) Automatic identification system (AIS): data reliability and human error implications. J Navig 60(03):373–389

Chapter 13
Machine-to-Machine Model for Water Resource Sharing in Smart Cities

Banica Bianca and Catalin Negru

Abstract Nowadays, the level of natural resources exploitation is more complex regarding both quantity and quantity perspectives; therefore, the concern for reducing waste and caring about environmental impact has raised proportionally. Water represents an essential resource for people's needs, but also for industry and agriculture. This makes it a priority when it comes to finding innovative methods for its efficient usage. Improving cloud schemes, remodeling IoT applications, and integrating different systems that were once thought to work independently are steps forward toward interoperability in Water Management. This paper discusses the current initiatives in water management, building an image on what needs are being served, and what small or big solutions are being implemented. The model proposed is a solution for the management of a specific scenario using existing tools which need to be integrated.

Keywords Machine-to-machine · Water · Resource management

13.1 Introduction

Over the last decades, our cities have been drastically changing, following up on the people's needs with technology aid and with infrastructure remodeling. Natural resources exploit level has become more complex regarding both quantity and quantity perspectives; therefore, the concern for reducing waste and caring about environmental impact has raised proportionally. Water is an everyday life essential resource for basic people's needs, but also industries and agriculture which makes it a priority when it comes to finding innovative methods for its proper management. The population being on a rise, but the water being a limited resource, several issues are

B. Bianca · C. Negru (✉)
University Politehnica of Bucharest, Bucharest, Romania
e-mail: catalin.negru@cs.pub.ro

B. Bianca
e-mail: bianca.9.banica@gmail.com

© Springer Nature Switzerland AG 2021
F. Pop and G. Neagu (eds.), *Big Data Platforms and Applications*,
Computer Communications and Networks,
https://doi.org/10.1007/978-3-030-38836-2_13

a matter of present public discussion: water quality, water waste, and water-related natural disasters.

Domestic water supplies depend directly on the quality of water which has to have a tolerable level of risk, otherwise influencing negatively consumption, hygiene, food preparation, and therefore public health. This calls for a permanent monitoring of water quality and a continuous improvement on water filtering.

Taking into consideration the threat of water shortages, another high priority would be reducing leakage (known as Non-Revenue Water) in water piping systems. These leakages consist of clean water which is lost before reaching its customer/consumer and which can happen because of a non-detected in time fractured pipe. Up to 40% of safe and clean water is lost because of leakages [1] in big cities. The aging water infrastructure is very often prone to defects. Although large diameter pipes are expensive to replace, the risk of rupture and further massive water leakage is very high. By monitoring pressure, flow volume, direction, and other conditions from within the city water networks (in more than a few key points situated kilometers away), actions can be taken before actual damage occurs and a better understanding of the water network will be provided.

Weather hazards that provoke natural disasters are 90% related [1] to water management issues. Sewerage city systems and water supplies need to be properly handled in order to avoid negative consequences which have more than once led to the loss of life, damage of properties and infrastructure, and human health issues. Preparation through collection and analysis of forecast information, relevant weather patterns, seismic activity, or climate change indicators is an essential part of modern water management requirements.

The most common natural disaster encountered in Europe is represented by floods (EEA, 2015). Usually, flooding occurs as a combination of phenomena, and the risk of it having a negative impact is amplified by the unsustainable urban/rural drainage systems. Mitigation against floods cannot be efficient without a water management system which takes into consideration water cycles, land planning integration with water factors, climate change, and weather conditions. New sustainable drainage systems and management techniques are being developed and implemented trying to promote these planning practices to all levels of government and industry, showing the benefits that these will have on both the environment and the communities.

Smart Environments grow out of the continuous integration of ICT (Information and communications technology) with everyday needs or with large-scale remodeling needs. This new arrival can provide the water management field a variety of architecture models in which data analysis results can be an essential support in decisions and actions of stakeholders, environmental associations, or simply residents. SCADA technologies which have been supporting business processes related to water industries are not enough in order to properly solve the water sustainability problem. Real-time data management or GIS (Geographical Information Systems) integrations are concepts which define new product/service design in water management area. Remote monitoring along with real-time diagnosis over report generation can optimize strategies of water distribution, water waste minimization, or water quality conservation.

The aim of this paper is to research and summarize the current initiatives in water management, to present the technologies used in the development and implementation of these initiatives, and to propose a solution for a small-scale water management problem or an improvement of an already running water management technology.

13.2   Current Stage of Development in the Field

13.2.1   EOMORES Project—Copernicus Platform

Earth Observation-Based Services For Monitoring And Reporting Of Ecological Status (EOMORES) is a water quality monitoring project initiated in 2017 by a group of researchers who were previously involved in different smaller initiatives of water quality data gathering such as FRESHMON (FP7, 2010–2013), GLaSS (FP7, 2013–2016), CoBiOS (FP7, 2011–2013), GLoboLakes (UK NERC, 2012–2018), and INFORM (FP7, 2014–2017). Many of these projects were funded by the European Union's Seventh Framework Programme for Research and Technological Development (FP7) and dealt with satellite data analysis in the monitoring of quality requirements imposed by the EU directives. The purpose of the EOMORES project is different scale monitoring on water bodies, combining a series of techniques in order to obtain comprehensive results which can be further used in an efficient manner.

The first technique is based on satellite monitoring and is providing a set of data from Copernicus Sentinels every few days, depending on the availability of the Sentinels or the weather conditions. The data provided is then converted into information by one or several algorithms which have been previously tested against a wide range of factors (location, type of water, and sensor technical characteristics). For the project research, EOMORES is collaborating with six countries (Italy, France, The Netherlands, the UK, Estonia, Lithuania, and Finland), having different latitude coordinates for their location and also belonging to very different climates/ecoregions.

The main focus of the observations was the comparison of different approaches to atmospheric correction ("the process of removing the effects of the atmosphere on the reflectance values of images taken by satellite or airborne sensors" [2]). Satellite data has the advantage of covering large areas, but the drawback is that they have limited levels of detail achievement and measurement frequency. EOMORES uses Sentinel-1, Sentinel-2, and Sentinel-3 from Copernicus programme which offers free and open data. The first launch of Earth Observation satellites for Copernicus happened in 2014 and one year after, they have improved their assets by launching Sentinel-2A which was aimed to provide "color vision" data for changes on the surface of the Earth [2]. This means that its optical system included three spectral bands in the electromagnetic spectrum area also called "red-edge" in which the difference is plant reflectance (visible light absorbed by plants versus radiation scattered in the photosynthesis).

The second technique consists of in situ observations or in situ monitoring which provides continuous measurements of a specific location in an assigned period of time (e.g., 24 h). This is information collected directly on-site and is meant to be complementary or validating of the satellite data. As this kind of observation is not influenced so much by the weather state, the level of control over the frequency of measurement is higher. EOMORES researchers use hand-held devices for in situ data collecting, but an autonomous fixed-position optical instrument is in development as an improvement of the current Water Insight Spectrometer (WISP). This instrument is used for measuring Chlorophyll, Phycocyanin, and Suspended sediments which are measures of algal biomass/cyanobacterial biomass/Total Suspended Matter (TSM) [3]. The measured values appear on the display in 30–90 s. The raw information can be uploaded from the WISP to a cloud platform for further analysis, model generation, and further computation. The WISP has incorporated three cameras that are able to break light into its spectral components. By comparing the three sources which measure the light coming straight into the WISP and the light reflected from the surface of the water, a derivation of the parameters wanted is being done through "band-ratio" algorithms.

The third technique is modeling which combines the results obtained through the two techniques previously presented in order to generate prediction data and forecast information on the specified area.

All the reliable quality water datasets are to be transformed into sustainable commercial services offered to international or national/regional authorities which are in charge of monitoring water quality or are responsible for water management and environmental reporting. Private entities that deal with the same monitoring issues can benefit from the data collected and mined by the EUMORES researchers.

13.2.2 AquaWatch Project

AquaWatch project (2017–2019) is part of the GEO (Group on Earth Observation) Water Quality Initiative whose aim is to build a global water quality information service. The implementation of the project, currently in progress, is based on activity distribution across working groups, each with a specific focus element.

AquaWatch has a targeted public which consists of science and industrial communities, non-governmental organizations, policymakers, environmental organization managers, and non-profit organizations. Also, access to information will be promoted to simple recreational users too. These potential end-users are to be attracted and involved as volunteers in the working groups or for gathering data. At the beginning of 2018, the first group of products should be finished. This group would contain products that support turbidity measurement using different techniques:

– a Secchi disk [4] depth product which is a plain, circular disk 30 cm in diameter for measurement of water transparency or turbidity in bodies in water;
– a diffuse attenuation coefficient product;

- a Nephelometric Turbidity Unit product;
- a surface reflectance product.

13.2.3 SmartWater4Europe Project

Smart Water For Europe is a demonstration project that is being created to produce business cases for Smart Water Networks (SWN). Founded by the European Union, the project will try to demonstrate optimal water networks and will look at the potential to integrate new smart water technologies across Europe. 21 European organizations are taking part in the project and four demonstration sites are available—Vitens (the Netherlands), Acciona (Spain), Thames (UK), and Lille (France).

The project will help to understand how small technologies can deliver cost-effective performance and improve the water supply service given to customers. Smart water technologies and hi-tech informatics will allow the early detection of leaks on a 24/7 basis leading toward Smart Networks through data capture, analysis, and reporting. As well as leakage detection, service excellence will be supported with best practices as energy monitoring, water quality, and customer engagement. The main idea of the project is to implement small projects, receive recommendations based on the implementation, and then go on with another small project in a different location, across Europe and beyond.

Vitens Innovation Playground (The Netherlands) is a demonstration site which consists of 2300 km of distribution network, serving around 200,000 households. Conductivity, temperature, and chlorination level are measured using hi-tech sensors. Pipe burst detection and water hammer detection are tested through Syrinix sensors. Using an integral ICT solution, all the dynamic data from sensors, but also static data formats such as area photos, distribution network plans, or soil maps are stored and made available for water companies participating in the project and for researchers. The Vitens Innovation Playground serves also as a training facility in which operators learn how to respond to high-risk incidents like contamination or massive leakages.

Smart Water Innovation Network in the city of Burgos (Spain) is working with three different hydraulic sectors (one industrial, one urban, and one residential) which have been converted into one Smart Water Network. To create it, a network of quality sensors and conventional water meters with electronic versions equipped with communication devices have been installed. The information provided by the sector flow meters is integrated into the end-to-end management system, the so-called Business Intelligence Platform. This platform, hosted in the Big Data Center, manages and processes the data gathered from common management systems, integrating also the algorithms developed in order to automatically detect leaks, predict consumption levels, and check the quality of the water at any moment. The platform information is also used for continuously improving the overall service. The ultimate objective is to find the key parameters of the smart supply network so that the service could be implemented in any location, regardless of its characteristics. Among the benefits of such projects is the chance of savings of millions of dollars all over the world.

Thames Water Demo Site (The Netherlands) focuses on trunk mains leak detection by being aware of transients or rapid changes in pipe pressure and taking proactive action about the specific incidents. In addition, a first attempt has been made to distinguish between customer side leakage and wastage through a scalable algorithm that has been trained on smart meter data. In order to promote good practices, customers have been given incentives to save money and earn discounts by using water more carefully. An energy visualization tool was built in order to show where the energy on the network is being distributed. This graphic tool helps users better understand the dependency between demand, pressure, and energy. All the solutions are concentrated in a single interface in order to display relevant information for operators to act or to discover cause–effect connections.

13.2.4   OPC UA with MEGA Model Architecture

The authors of [5] identify the problem of interoperability in water management initiatives, caused by the lack of support and lack of standardization in the monitoring processes, as well as the control equipment. They propose a smart water management model which combines Internet of Things technologies and business coordination for having better outcomes in decision support systems. Their model is based on the OPC US (Object Linking and Embedding for Process Control Unified Architecture) which is an independent platform that offers service-oriented possibilities of architecture schemes for controlling processes which are part of the manufacturing or logistic fields. The platform is based on web service technologies, therefore being more flexible to scenarios of usage.

The proposed model is the MEGA model which takes into consideration functional decoupled architectures in order to achieve the goal of increased interoperability between the water management solutions on which companies and organizations are currently working. This would also solve the problem of SME (Small and Medium-sized Enterprise) companies locally oriented which provide good local solutions for water management, but which have difficulties in expanding to other countries or regions, or to maintain their funding in a long term.

The MEGA architecture consists of several layers, the main ones being the following:

– Management and Exploitation layer—hosts the main applications and services (can be executed in cloud, on local hosts) and supports the management definitions of the processes;
– Coordination layer—defines and can associate, if necessary, entities to physical objects, collects the procedures defined by the ME layer, and delivers them to the Subsystem layer after associating sequence of activities to them (recipes);
– Subsystem layer—contains the subsystems that execute, independently or not, the procedures and recipes defined in the Coordination layer;

– Administration layer—provides a user interface for administration and monitoring and enables configuration of entities defined in previous layers.

The water management model proposed includes a Physical Model and a Process Model which contain several Process, Cells, Units, Units Procedures, Control Modules, Equipment Modules, and Operations which can be handled differently, according to the business requirements. The big steps of the whole Mega Model process are as follows:

– Identifiers Mapping—map recipe identifier to subsystem identifier (if the recipe is already provided, if not, translate the instructions into a standard recipe first);
– Recipe validation—check if the subsystem is able to execute the process contained in the recipe;
– Process transfer to the suitable subsystem—each subsystem receives its sequence of activities to be executed;
– Control and monitoring of the process execution—information about the on going processes can be monitored in real time.

13.2.5 WATER-M Project

WATER-M project is an international initiative of representatives from four countries (Finland, France, Romania, and Turkey), part of the Smart City challenge. The project is meant to contribute to a major upgrade of the water industry by helping with the introduction and integration of novel concepts such as GIS (Geographic Information System) usage, ICT with IoT applications, or real-time data management or monitoring. The final purpose is to build a unified water business model targeted at European Union water stakeholders. Through operational control and monitoring real-time data, the WATER-M project is currently developing a service-oriented approach and event-driven mechanisms for dealing with the water sustainability problem.

As the project was started in 2017, the plans and results are made public once progress is made. The use cases defined for this initiative are stated below [6]:

– Leak Detection;
– Development of Water Management and Flood Risk Prevention Platform;
– River Tele-monitoring;
– Performance monitoring of water distribution network;
– Control and optimization of the water distribution network;
– Coordinated management of networks and sanitation structures;
– New redox monitoring;
– Urban Farming.

Energy cost reduction and compatibility with European directives on the water for allowing new business models for water management to emerge on the basic structure of the WATER-M are taken into consideration. Critical challenges, as well as options for various communication protocols such as LTE-M or LoRa, or AMR (Automatic Meter Reading) technologies with benefits and drawbacks were discussed in a state-of-the-art [7] aimed at evaluating the previous proposals in the areas of water management. A new model has not yet been proposed, it is still a work in progress.

13.3 Smart City Water Management Available Technologies

13.3.1 GIS (Geographic Information System)

A GIS system can be viewed as a database, which comprises all geometric elements of the geographical space with specific geometric accuracy together with information, i.e., in tabular form which is related to geographic location. The GIS is associated with a set of tools, which do data management, processing, analysis, and presentation of results for information and related geographic locations. The geographical space can be viewed as composed of overlaid planes of information over a wider geographical area and each plane has specific information or features [8].

The different planes contain similar geographic features. For example, one plane has elevations, another plane can have the drainage features represented, while another can have the rainfall. Thematic maps are then created, using map algebra on plane information [9–11].

All the features in GIS are viewed as objects which can further be used to build models. The simplest object is a point object and the complex/composed objects such as lines or areas rely on the point objects.

The up-to-date GIS technology is able to use data stored in warehouses or databases, accessing it through the internet and running the GIS system every time the specific datasets change. This is a feature usually used in order to have reliable real-time hydrological models for forecasting systems. Further developments on GIS technology are aimed at integrating object-oriented programming techniques, therefore ordering components into classes. An example of a component may be a line segment of a river and the data contained in such a class can represent coordinates, length values, profile dimensions, or procedures for computing the river flow at a specific moment.

Water management could use GIS systems for basic data such as creating a national hydrology dataset which is permanently updated, but also for hydrologic derivatives which can be used together with satellite data and in situ information for dealing with prevention and management of water shortage or better organizing cities and rural areas.

13.3.2   IBM Water Management Platform

IBM Water Management Platform is a Big Data Cloud platform offered by IBM for implementation of solutions which can help end-users or organizations in several forms, regarding environmental or direct water problems. The set of features offered by the problem can be summarized as follows: (1) provide situational awareness of operations, (2) integrate data from almost any kind of source (GIS, ERP-Enterprise Resource Planning, satellite, on-site data-photo, video, numerical), (3) form patterns and correlations, visualize graphically contextual relationships between systems, (4) run and monitor SOPs (Standard Operation Procedures) from dashboards, (5) no compatibility adjustments needed when adding or removing devices, (6) set up business rules for generating alerts in risky situations, and (7) compare current and historical data to discover patterns or cause–effect relations.

IBM Intelligent Water solutions offer multiple deployment models to provide options for cities of all sizes with varying levels of IT resources. Cities with robust IT capabilities or strong interests in "behind-the-firewall" implementation can deploy this solution in their own data centers. Alternatively, deploying IBM Intelligent Water on the IBM SmartCloud can help cities capitalize on the latest technological advances while controlling costs [12].

Also, the personalized views are used by different so-called role-given-users for efficient analysis. The platform offers Citizen View (for water track usage in households), Operator View (for events, assets on geospatial maps), Supervisor View (for trends against KPI-key performance indicators), and Executive View (tracking and communicating KPI updates).

IBM Intelligent Water products are currently used in the Digital Delta system in the Netherlands which analyzes data in order to forecast and prevent floods in the country, while the city of Dubuque (United States) uses the IBM platform for sustainable solutions in household water consumption, monitoring infrastructure leakages and reducing water waste.

13.3.3   TEMBOO Platform—IoT Applications

Temboo is a software toolkit available directly from the web browser which enables anyone to access hard technologies like APIs (Application Programming Interfaces) and IoT (Internet of Things). Temboo users have access to data through public and private APIs and can develop their own IoT applications, starting from the services offered by the platform.

Developers would use what Temboo calls "choreos" to build together an application that is triggered from inputs registering on the IoT ARTIK device. Choreos are built out of APIs and act like microservices that perform one specific function that might be made available through an API. By splitting an API's functionalities into microservices using the choreos format, code snippets can be kept short and memory requirements and processing power can be reduced on the device itself, while also enabling a more complex server-side processing to be undertaken in the cloud [13]. Hardware development kits, embedded chipsets, sensors and data from sensors, actuators and remote control of actuators, M2M communication frameworks, and gateway/edge architectures can be integrated into Temboo. It generates editable pieces of software code which is in a standardized form, partitioned into production-ready blocks, and easy to implement with the aid of cloud services.

Temboo offers lightweight SDKs, libraries, and small-footprint agents for programming every component: MCUs (C SDK/Library, Java Embedded (in progress)), SoCs/gateways (Python Agent with MCU, Java Agent with MCU, Python SDK, Java SDK), Mobile Applications (iOS SDK, Android SDK, Javascript SDK). For connecting devices to the cloud services, Temboo supports Bluetooth, Ethernet, WiFi, and GSM (in progress).

Temboo can generate code for complete multi-device application scenarios, in which edge devices use a common IoT communications protocol to send Temboo requests through a gateway. The gateway handles all communication with Temboo, enabling local edge devices to interact with the huge range of web-based resources supported by Temboo [14]. The protocols used for M2M (Machine-to-Machine) communications are MQTT, CoAP, or HTTP.

Message Queuing Telemetry Transport (MQTT) is a standard for publish–subscribe-based messaging protocols. It works on top of the TCP/IP protocol and is used for connections with remote locations with constraints for network bandwidth [15].

Constrained Application Protocol (CoAP) [16] is a service layer protocol in well-suited internet devices, such as wireless sensor network nodes which are resource limited. This protocol enables nodes to communicate through the Internet using similar protocols. It is also used with other mechanisms, like SMS on mobile communication networks.

A series of pre-build applications are provided which are demonstrated on a small scale, but can be also used for large-scale problems. One of those applications is Water Management for monitoring and remotely controlling the water level in a tank. This includes a mobile alert sent to the user in case of action needed to be taken on the water tank level.

13.3.4 RoboMQ

RoboMQ is a Message Queue as a Service platform hosted on the cloud and also available as an Enterprise hosting option. This Software as a Service (SaaS) platform is an integrated message queue hub, analytics engine, management console, dashboard and monitoring, and alerts; all managed and hosted in a secure, reliable, and redundant infrastructure" [17].

The key features that the platform offers are: (1) Scalability (auto-scalable through any load balancing and scaling), (2) Expandability (it can be integrated in application or other features/functions can be added to it), (3) Reliability (messages are persistent and durable), (4) Monitoring through dashboards, analytic tools, and specific alerts, (5) Compatible with different protocols (MQTT, AMQP (Advanced Message Queueing Protocol), STOMP (Simple Text-Oriented Messaging Protocol), HTTP/REST), (6) Support for multiple programming languages (all the libraries supporting the protocols above are supported by RoboMQ (e.g., Python, Java, NET), (7) Secured connections (support SSL (secure socket layer) connection for all available protocols).

RoboMQ acts as a message broker, managing queues between a producer and a consumer. Given its expandability feature, it has been integrated into an IoT Analytics application which collects data from various sensors and sends it to the queues managed by RoboMQ. The data is redirected to an IoT listener which then writes in a specific real-time database. All the data can be monitored through dashboards, panel metrics, and graphs in real time.

RoboMQ provides M2M integration through an open standard-based platform to connect devices and sensors to the back-end applications, systems, or processes. The protocols supported by RoboMQ (MQTT, STOMP, AMQP) can run on very small-footprint devices using one of the languages that are supported by the device OS and profile. Among the devices that can be used are: Raspberry Pi, Audrino, Beaglebone, and mBed-based platforms. These devices will have the role of producer that send the data as messages through to the RoboMQ broker, while the consumer will be the RoboMq dashboard application.

13.4 Proposed Model and Possible Directions

Taking into consideration the possibilities offered by the ICT technologies and the critical problems in the water management field, a model of M2M device collaboration is proposed. The main purpose is optimization of water resource sharing. This model represents an M2M integration between RoboMQ (message broker) and Temboo (IoT software toolkit) to coordinate the distribution of the same available water resource when several requests are made at the same time. We use the following methods:

– Use labeled queues to differentiate between messages (data values), therefore evaluating the greater need before sending the commands to the actuators;
– Tune parameters for obtaining a generic water-saving mode which the user can set when receiving several notification alerts of water shortage (expand for usage on large scale, e.g., city scale) Targeted at/ Use Cases;
– Regular end-users for better management of household or small facilities water resources—farms, rural houses, residences with their own water supply, and zoo/botanic gardens;

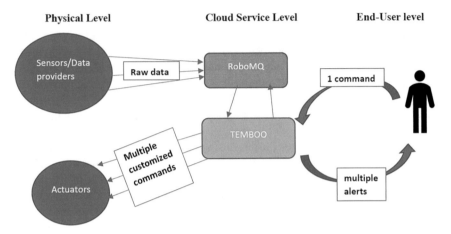

Fig. 13.1 Proposed architecture

– Authorities for better management of single city water resources in critical situations—prolonged water shortage, prolonged repairs to the water infrastructure, and natural disasters.

The architecture of the proposed model is presented in Fig. 13.1. The architecture is structured on three levels: Physical level, Cloud Service level, and End-User level. At the Physical level exist sensors that transmit raw data to a RoboMq service, and actuators that receive multiple customized commands from a TEMBOO service. At Cloud Service level exist two systems ROBOMQ that receive data from sensors and TEMBO that send commands at the physical level. The top level in End-User level takes as input commands from users and receives multiple alerts from physical and cloud service level.

13.5 Possibilities of Implementation

In order to be able to build a solution for the M2M model proposed, the elements needed in the integrations have to be identified. The intention is to integrate two different entities, one being a system of sensors and actuators and the other one a mobile/desktop application that offers the possibility of receiving a notification/message alert but also of giving back a response. The communication between the two systems, or better said, between the system and the end-user can be done through a Message-Oriented Middleware (MoM), while the flows of action can be implemented into microservices (e.g., email alert microservice).

RabbitMQ has been chosen as a MoM for a performance analysis in order to confirm if this type of middleware is suitable for the model proposed.

13.5.1 Message-Oriented Middleware—RabbitMQ

RabbitMQ is a message-queueing software usually known as a message broker or as a queue manager. It allows the user to define queues to which applications may connect and transfer messages, along with the other various parameters involved.

A message broker like RabbitMQ can act as a middleman for a series of services (e.g., web application in order to reduce loads and delivery times). Therefore, tasks that would normally take a long time to process can be delegated to a third party whose only job is to perform them. Message queueing allows web servers to respond to requests in a quick way, instead of being forced to perform resource-heavy procedures on the spot. Message queueing can be considered a good alternative for distributing a message to multiple recipients, for consumption, or for balancing loads between workers.

The basic architecture of a message queue is based on several elements: client applications called producers that create messages and deliver them to the broker (the message queue), and other applications called consumers that connect to the queue and subscribe to the messages. Messages placed in the queue are stored until the consumer retrieves them.

Any message can include any kind of information. It could have information about a process that should start on another application (e.g., log message) or it could be just a simple text message. The receiving application processes the message in an appropriate manner after retrieving it from the queue. Messages are not published directly to a queue, but, the producer sends messages to an exchange that is responsible for the routing of the message to different queues. The exchange routes the messages to message queues with the help of bindings (link) and routing keys [18].

13.5.1.1 Test Performance on RabbitMQ

The aim of this test is to assess/analyze the performance of RabbitMQ server under certain imposed conditions. In order to run the tests, a CloudAMQP instance hosting RabbitMQ solution will be used. RabbitMQ provides a web UI for management and monitoring of RabbitMQ server. The RabbitMQ management interface is enabled by default in CloudAMQP.

Parameters monitored: Queue load, Publish message rate, Delivery message rate, Acknowledge message rate, Execution time, Lost messages, and Memory usage as can been seen in Fig. 13.2.

We can draw the following observations based on results:

– When having only one consumer, the Queue load value increases proportionally with the number of messages sent (n-30k → 2n-60k → 3n-80k);
– Queue load is reduced by approximately 15% when increasing the number of consumers from 1 to 10;
– Execution time is reduced by approximately 70;

	Scenario 1				Scenario 2				Scenario 3			
	n	2n	3n	n(long)	n	2n	3n	n(long)	n	2n	3n	n(long)
Queue load	30k	60k	80k	20k	28k	47k	83k	10.2k	17.5k	45k	62k	6
Execution time	100min	120min	180min	105min	35min	75 min	85min	38min	5min	10min	19min	25min
Publish rate	82/s	150/s	50/s	38/s	240/s	100/s	420/s	26/s	260/s	300/s	310/s	23/s
Delivery rate	9/s	9/s	9/s	7/s	19/s	20/s	18/s	18/s	91/s	95/s	98/s	23/s
Acknowledge rate	9/s	9/s	9/s	7/s	19/s	20/s	18/s	18/s	91/s	95/s	98/s	23/s
Lost messages	0	0	0	0	0	0	0	0	0	0	0	0

	Scenario 4								Scenario 5							
	before disconnect of 1 worker				after disconnect of 1 worker				before disconnect of 9 workers				after disconnect of 9 workers			
	n	2n	3n	n(long)	n	2n	3n	n(long)	n	2n	3n	n(long)	n	2n	3n	n(long)
Queue load	-	40k	-	-	20k	-	60k	4	23k	32k	-	-	-	-	72k	20k
Execution time	-	-	-	-	7min	12min	19min	20min	-	-	-	-	40 min	70 min	177min	65min
Publish rate	200/s	270/s	351/s	31/s	236/s	0/s	258/s	30/s	320/s	278/s	181/s	23/s	0/s	0/s	326/s	31/s
Delivery rate	90/s	93/s	91/s	31/s	85/s	79/s	83/s	30/s	92/s	93/s	78/s	23/s	9.4/s	9,4/s	9.2/s	7.4/s
Acknowledge rate	90/s	93/s	91/s	31/s	85/s	93/s	83/s	30/s	92/s	93/s	78/s	23/s	9.4/s	9,4/s	9.2/s	7.4/s
Lost messages	0	0	0	0	0	0	0	0	0	0	0	0	0	0	0	0

Fig. 13.2 RabbitMQ test scenarios

- Publish rate is directly influenced by the size of the message being sent, but is independent of delivery rate;
- Queue load is directly influenced by the delivery rate;
- Delivery rate is increased proportionally with the number of consumers, when sending a short message (1 consumer-9/s → 2 consumers-18/s → 10 consumers-91/s);
- When sending long messages and having multiple consumers, publish rate and delivery rate have close values, hence queue load is very small. The time needed for the message to be published is almost the same as the time needed for the message to be sent and acknowledged;
- When sending long messages and having one consumer, the same theory as in the short message case is applied, queue load is remarkably increased (6 to 20k), and delivery rate is lowered to a value smaller than the rate per user receiving short messages (30/s to 7/s);
- When killing one or multiple consumers in the send/receive process, the messages are redirected to the other running consumers. No other messages are lost, except for the ones that were already acknowledged by the consumer which was disconnected;
- Messages are not equally distributed to multiple consumers, but the values are similar enough (e.g., for 10 workers: 3030, 3005, 3014, 2966, 2998, 2995, 3023, 2978, 2988, 3002);
- When having a send/receive process without acknowledgment, queue load is 0, as the messages are continuously sent, without waiting for a response from the consumer. This approach is risky, as the user has no information about possible lost messages.

RabbitMQ offers an efficient solution for message queuing, which is easy to configure and integrate into more complex systems/workflows. It can withstand and successfully pass stress load bigger than 10k calls and it decouples front-end from the back-end.

The most common disadvantage is related to troubleshooting, as users have no access to the actual routing data process. A graphical interface or access to inner parameters would be useful when dealing with large clusters.

13.6 Conclusions

As smart cities emerge, new solutions are needed for every resource management out there. Water management is an area that needs solutions, both for regular basis usage and for critical situations. Since technology has become more and more sophisticated, but at the same time more user-friendly, opportunities for developing ideas with easy-to-understand tools have appeared.

Although on a large, up to global scale, the focus is on standardization of water management processes and building business models that can be feasible regardless of the conditions/location, pilot solutions on a local/little scale are the ones that support the larges research, through beneficial continuous small results or continuous tries. Improving cloud schemes, remodeling IoT applications, and integrating different systems that were once thought to work independently are steps forward toward interoperability in Water Management.

This paper discussed the current initiatives in water management, building an image on what needs are being served, and what small or big solutions are being implemented. The model proposed a solution for the management of a specific scenario using existing tools which need to be integrated. The second part of the paper will contain possibilities of implementation and case studies on the proposed model.

Acknowledgements Research was supported by UEFISCDI, through the PN III project no. 16/2016, Awarding Participation in H2020—Data4Water, no. 690900.

References

1. GSMA Homepage. https://www.gsma.com/iot/wp-content/uploads/2016/11/Smart-water-management-guide-digital.pdf. Last accessed 7 January 2018
2. COPERNICUS Homepage. http://newsletter.copernicus.eu/issue-11-september-2015/article/launch-sentinel-2a-brings-colour-vision-copernicus-programme. Last accessed 7 January 2018
3. EOMORES-H2020 Homepage. http://eomores-h2020.eu/blog/what-is-a-wisp-anyway/. Last accessed 7 January 2018
4. Preisendorfer R (1986) W: Secchi disk science: visual optics of natural waters1. Limnol Oceanogr 31(5):909–926

5. Robles T, Alcarria R, de Andrés DM, Navarro M, Calero R, Iglesias S, López M (2015) An IoT based reference architecture for smart water management processes. JoWUA 6(1):4–23
6. ITEA3 Homepage. https://goo.gl/CC61Q8. Last accessed 7 Jan 2018
7. Gebremedhin B (2015) Smart water measurements: literature review. Water-M Project, 13 June 2015
8. Hatzopoulos JN (2002) Geographic information systems (GIS) in water management. In: Proceedings of the 3rd international forum integrated water management: the key to sustainable water resources
9. Gorgan D, Bacu V, Rodila D, Pop F, Petcu D (2010) Experiments on ESIP–environment oriented satellite data processing platform. Earth Sci Inf 3(4):297–308
10. Petcu D, Zaharie D, Gorgan D, Pop F, Tudor D (2007) MedioGrid: a grid-based platform for satellite image processing. In: 4th IEEE workshop on intelligent data acquisition and advanced computing systems: technology and applications, 2007, IDAACS 2007. IEEE, pp 137–142
11. Pop F (2007) Distributed algorithm for change detection in satellite images for grid environments. In: Sixth international symposium on parallel and distributed computing, 2007, ISPDC'07. IEEE, pp 41–41
12. IBM Intelligent Water, infrastructure Services Documentation, IBM Industry Solutions-Solution Brief. https://www-935.ibm.com/services/multimedia/Intelligent_Water.pdf. Last accessed 7 January 2018
13. Mark Boys, Temboo API Platform Puts Industrial IoT in Reach. https://www.programmableweb.com/news/temboo-api-platform-puts-industrial-iot-reach-devs/analysis/2015/05/28. Last accessed 7 January 2018
14. TEMBOO Homepage. https://temboo.com/hardware/m2m-mqtt. Last accessed 7 January 2018
15. Hunkeler U, Truong HL, Stanford-Clark A (2008) MQTT-S—a publish/subscribe protocol for wireless sensor networks. In: 3rd international conference on communication systems software and middleware and workshops, 2008, COMSWARE 2008. IEEE
16. Shelby Z, Hartke K, Bormann C (2014) The constrained application protocol (CoAP)
17. ROBOMQ Homepage. https://robomq.readthedocs.io/en/latest/. Last accessed 7 January 2018
18. Website documentation. https://www.cloudamqp.com/blog/2015-05-18-part1-rabbitmq-for-beginnerswhat-is-rabbitmq.html. Last accessed 3 June 2018

Index

A

Ad hoc network, 84, 96, 100
Administration, 71, 117, 127, 239, 244, 277
AES, 211, 213, 215–218
Agent implementation, 100
Aids to navigation, 256, 257
Alert service, 192–194, 198
Android, 113, 155–159, 216, 217, 219, 280
AquaWatch, 274
Arduino, 109, 228–230
Asynchronous interviewing, 61
Attacks, 114, 120, 122–124, 128, 134, 136, 139, 142, 143, 145, 154, 162, 185
Automatic Identification System, 253, 255, 260, 268, 269
Automatic assessment, 181, 192–194, 198, 199, 202, 205
Availability, 4, 6, 30, 32, 88, 117, 120, 146, 147, 153, 154, 168, 173, 176, 178, 189, 239, 255, 273
Awareness, 4, 6, 12, 15, 70, 72, 99, 168, 182, 209, 279

B

Batch processing, 82, 84
Battery life, 139, 148, 154, 226
Big data, 1–5, 10, 32, 37, 39, 51, 54, 78–82, 84–86, 165, 166, 172, 177, 238, 268, 275, 279
Bio-inspired techniques, 39, 41
Broker, 60, 63, 66, 120, 124–127, 130–132, 134–137, 140–145, 156, 238, 281, 283

C

Cargo tracking, 257
Ciphertext, 144, 208, 210, 211
Client, 63, 94, 102, 118–120, 126–130, 132, 136, 138–140, 142, 143, 150, 157, 158, 188, 192, 196, 199, 207, 208, 283, 243–246
Client-server, 119, 124, 127–129, 139, 142, 144, 145, 199
Cloud computing, 39, 87, 88, 113, 116, 121, 175, 208, 210, 227, 237–239
Cloud security, 114
Cluster, 1–4, 7–10, 12–18, 23, 24, 26–30, 32, 33, 82, 285
Columnar formats, 78, 79
Communications, 39, 60–63, 65, 66, 68, 70, 72, 84, 88, 89, 94, 99, 102, 116, 117, 119–128, 131, 132, 135–137, 139, 140, 142, 143, 146, 149, 150, 152, 154–156, 159, 161–163, 166, 167, 169, 170, 172, 173, 178, 183, 184, 191, 197, 199–201, 209, 219, 220, 227, 228, 233, 239, 269, 272, 275, 278, 280, 282
Confidentiality, 116, 123, 126, 130, 134, 136, 155, 207, 210, 212, 215–218
Constrained Application Protocol, 119, 280
Copernicus, 273

D

Data acquisition, 24, 84, 181, 189, 191, 194–196, 204, 205
Data analysis, 3, 9, 10, 13, 15, 16, 23, 62, 84, 121, 255, 258, 261, 268, 269, 272, 273

© Springer Nature Switzerland AG 2021
F. Pop and G. Neagu (eds.), *Big Data Platforms and Applications*,
Computer Communications and Networks,
https://doi.org/10.1007/978-3-030-38836-2

Data center, 1, 2, 5, 7, 8, 15, 17, 30, 33, 86, 87, 208, 279
Data collection, 3, 39, 62, 82, 121, 167, 168, 183, 193
Dataloggers, 184
Data-photo, 279
Data presentation, 168
Data processing, 52, 113, 115, 116, 118, 120, 121, 157, 162, 168, 174, 183, 189, 190, 192, 220
Data reduction, 253, 254, 261, 263
Data storage, 181, 183, 192, 193, 205
Data visualization, 264
Decision system, 193, 200, 202
Deep learning, 37, 38, 40, 41, 49, 86
Delay tolerant workload, 6
Design patterns, 113–117, 121–124, 127, 129–132, 134–140, 142–150, 152–157, 159–163
Dew computing, 169–171, 176, 178
Digital certificates, 125, 126, 137, 143
Discrete manufacturing processes, 37, 38, 42, 44, 45, 51, 54
Drop Computing, 87–89
DSA, 214
Dynamic information, 261

E
Earth observation, 273, 274
Edge computing, 4, 5, 33, 87, 88, 170, 171, 176, 178
Education, 4, 59, 61, 71, 238, 239
E-mail interviews, 61, 62, 68
Encryption, 114, 144, 145, 207–213, 215–221
Energy efficiency, 1–7, 13, 15, 20, 28–33, 225, 229, 231, 234
Energy metrics, 12
Energy waste, 1, 4, 6–8, 12–15, 17, 18, 20–26, 28, 29, 32, 33
ESP8266, 227, 228
Exhibitions, 91, 92, 94–105, 107, 108

F
First Come First Served algorithm, 9
Fishing fleet monitoring, 257
FlashAir, 99–102
Fleet tracking, 257
Flink, 82–84
Fog computing, 87, 88, 237

G
Generic Modeling Environment, 240
Geolocation, 97, 99, 101, 102, 105, 107
Google File System, 82
GPS, 100, 192, 203, 228, 254
GSM, 191, 197, 280

H
Hadoop, 79, 80, 82–84, 86
Hadoop Distributed File System, 82
Hash functions, 214, 220, 221
Healthcare, 4, 165–168, 176–178
Heterogeneous data, 32, 77, 78, 81, 89, 254
High Performance Computing, 85, 87, 239
Home-based telecommuting, 64
HTTP, 119, 122, 127–129, 131, 133, 134, 137–139, 150, 196, 243, 244, 280, 281

I
IBM Water Management Platform, 279
IBM Service Delivery Manager, 237, 242
Incremental learning, 255
Information exchange, 102, 165, 166, 168
INSPIRE, 187, 188, 192, 194, 203
Interoperability, 10, 115, 117, 121, 146, 151, 187, 189, 192, 216, 239, 271, 276, 285
IOS, 217, 280
Internet of Things (IOT), 2, 4, 32, 39, 54, 87, 91, 93, 94, 96, 101, 113–163, 165, 166, 169–178, 225–227, 234, 271, 276, 277, 279–281, 285

K
Kafka, 43, 80, 83, 84
KPI, 279

M
Machine learning, 37–39, 41, 42, 44, 45, 49–51, 84–87, 115, 255
Machine-to-machine, 280
Management, 3, 4, 6–8, 14, 28, 30–33, 39, 63, 67, 72, 73, 79, 81, 83, 115–120, 126, 128, 137, 146–148, 150, 152, 155–160, 162, 163, 166, 168, 181, 182, 188, 201, 204, 208, 210, 215, 218, 242–246, 255, 271–283, 285
Maritime security, 257
MC++, 260
Medical Devices, 78, 167, 175

Memory-rich implementation, 99, 101, 103, 104
Message integrity, 136
Metadata, 105, 106, 145, 187, 220, 221, 229, 255, 262
Metering system, 182, 204
MHealth, 165–178
Mobile cloud computing, 87–89, 227
Mobile privacy, 88
Mobile work, 60, 63, 65, 67
Mobility, 94, 97, 99, 104, 117, 121, 239
Modeling as a Service, 238, 240, 242, 249
Modeling languages, 237, 238, 240, 242, 250
Monitoring, 2–5, 8, 13, 15, 17, 27, 30, 37, 51, 89, 149, 150, 153–155, 165–168, 170, 175, 177, 178, 181–194, 198, 204, 205, 237, 238, 243, 245, 254, 255, 268, 272–277, 279–281, 283
Moving objects, 94, 101
MQTT, 119, 122, 124–127, 130–133, 135–137, 140–145, 280, 281
MySQL database, 43

N
Neighbourhood Work Centres, 63, 65
Networking, 68, 87, 121, 122, 160, 167, 173
Network protocol, 113, 116, 122, 134, 137, 138
NSF
NoSQL database, 79, 81
Numerical methods, 87

O
Optimization, 5, 6, 31, 37–42, 47, 48, 51–54, 129, 156, 277, 281
Overlays, 208, 209, 215–219

P
PostGIS, 188, 261
Prediction, 31, 45–47, 49, 51, 158, 186, 261, 262, 266, 268, 274
Prevention, 166, 168, 277, 278
Privacy, 88, 144, 207–209, 215
Privacy enhancements, 207
Processing models, 81, 84
Proxy server, 128
Publish-subscribe, 114, 117, 119, 120, 124, 130, 136, 140, 141, 144, 238, 280

R
RADAR, 78, 254

Raw data, 78, 167, 168, 174, 183, 192, 255, 282
Real-time processing, 82, 83
Real time workload, 6
Redis database, 141
Relational database, 79
Resource management, 30, 31, 285
RFID, 97–99, 101–105, 109
RoboMQ, 280–282
RSA, 125, 162, 214, 216
RSS, 184

S
Samza, 82, 83
Satellite offices, 63–65
SCADA, 272
Scalability, 79, 82, 83, 88, 101, 115, 117, 121, 126, 146, 280
Scheduling, 6, 8, 9, 18, 30, 31, 33, 62, 145
Search and Rescue, 256, 257
Security, 4, 7, 67, 88, 97–100, 102–104, 113–131, 133, 134, 136, 137, 139–144, 146–155, 157–163, 165, 207–210, 212–216, 218
Self-organization, 94, 96
Semi-structured data, 79
Sensor Cloud, 238
Sensor network, 182–186, 189, 194, 195, 237–242, 249, 250
Sensors, 37, 38, 43, 45, 51, 77, 78, 87, 91, 101, 105, 107, 109, 121, 122, 128, 134–136, 148, 151, 154, 157, 167–170, 174, 175, 182–198, 200, 204, 227–230, 232–234, 237–242, 249, 250, 253, 273, 275, 280–282
Sentinel, 273
Server, 3, 24, 31, 43, 78, 94, 97–101, 103, 114, 119, 120, 126–129, 131, 133–135, 138, 139, 158, 160, 167–170, 172–176, 188, 192–197, 199, 203, 212, 215, 221, 228, 229, 237, 242–244, 248, 280, 283
Service Oriented Architecture, 238
Similarly, 24, 85, 92, 94, 98, 255, 268
Simulation, 15, 16, 24, 25, 39, 41, 43, 192–194, 198, 202, 203, 205, 240
Smart pocket devices, 170
Smart Cities, 1–6, 15, 24, 27, 28, 32, 33, 113, 114, 157, 277, 278, 285
Software, 3, 5, 9, 24, 28, 43, 100, 102–105, 113–119, 125, 127, 145, 148, 150, 153, 156, 157, 161, 162, 167, 171,

173, 175, 176, 185, 207, 215, 217–219, 238, 239, 243, 245, 246, 249, 279–281, 283
Spark, 37, 80, 82–84, 86
Sparse matrix, 12, 86, 87
Standard Operation Procedures, 279
Static information, 261
Storage System, 8, 82, 244, 254
Storm, 82, 83
Streaming, 84, 85, 166, 172–178
Stream processing, 81–84
Street traffic, 225
Structured data, 78, 79, 109, 269
Sustainability, 1–8, 20, 27, 28, 32, 272, 277
Symmetric cryptographic, 125, 126, 158
Synchronous interviewing, 61

T
Telecommuting, 59–65, 71
Teleworking, 59–63, 66–68, 71–73
Time-division multiple access, 256
TLS certificates, 140
Trust pattern, 131–133

U
Unstructured data, 78, 79
Useful work, 1–3, 7, 8, 12–15, 19, 20, 22–24, 26–28, 32, 33

V
Vessel traffic services, 254, 257
Video compression, 259, 260, 268
Virtual machine, 24, 145, 237, 243, 247–250

W
Water Insight Spectrometer, 274
Water resources, 182, 201, 204, 281, 282, 285
Water quality, 181–187, 189–194, 198–201, 203–205, 272–275
Wearable devices, 165, 167, 170, 172, 173, 176
Web application, 119, 159, 170, 171, 173, 174, 193, 194, 203, 204, 283
Wellness, 168
Wi-Fi, 97, 99–102, 104, 105, 107, 108, 174, 227, 228, 280
Wireless sensor networks, 183, 185, 189, 194, 195, 238, 280
Workload Management, 8

Z
Zabbix dataset, 9
Zookeeper, 42

Printed in the United States
by Baker & Taylor Publisher Services